Refreshingly, this lucid and engaging book devotes little time to handwringing over this "planetary crisis" and instead moves swiftly
systems to ensure that humanity lives within Earth's
views legal systems as complex, adaptive and highly ir
argues that a transition from environmental laws to ecc
bringing about what Thomas Berry referred to as a "m
Earth relationship." Geoffrey Garver is a leading ecol
book demonstrates why. By showing how to combine a
points" which can be used to create systemic change with an assessment of those features that "lock in" resistance to change, he makes an important contribution to our understanding of how to bring about the legal transformations necessary to establish ecologically viable human societies.

Cormac Cullinan, author of Wild Law: A Manifesto for Earth Justice

The timing of this publication is extremely fortuitous. Prior to the Corona-pandemic, ecological law might have been viewed as utopian, now it becomes vitally important for humanity and the book shows why. A precise analysis of systemic failure, it defines the principles and rules of Earth-based law and governance. A truly outstanding read.

Klaus Bosselmann, University of Auckland, New Zealand

Geoffrey Garver stands without peer on the intersection of ecological law and systems thinking. This nuanced and sophisticated book provides one of the clearest accounts yet about how law can be reconfigured to promote a mutually enhancing human-Earth relationship.

Peter D. Burdon, Associate Dean (learning and teaching), University of Adelaide Law School, Australia

This book offers a bold vision of ecological law, a radical shift of how we think and go about environmental regulation. It will be useful both for students and scholars of environmental law.

Giorgos Kallis, Universitat Autònoma de Barcelona, Spain

Absolutely indispensable, this fiery legal work introduces the content of ecological law, a conception of law of which the fundamental idea is to regulate the human-Earth relationship within ecological limits without neglecting law's major principle: justice. Geoffrey Garver is a pioneer and revolutionizes legal theory with his transdisciplinary approach and his direct proposal to test Ecological Law on the political economic level as a strategy for degrowth.

Cristiane Derani, Dean of Graduate Studies, Universidade Federal de Santa Catarina, Brazil

Ecological Law and the Planetary Crisis

This book uses a transdisciplinary systems approach to examine how Earth's human-caused ecological crisis arose and presents a new legal approach for overcoming it.

Ecological Law and the Planetary Crisis first examines how the history of humanity's social metabolism, along with the history of human inventions and ideas, led to the human-Earth dilemma we see today and explains why contemporary law is inadequate for confronting this dilemma. The book goes on to propose ecological law—law that maintains human activity within ecological limits such as planetary boundaries while ensuring social justice and equity—as an essential element of an urgently needed radical pathway of change toward a perpetual, mutually enhancing human-Earth relationship. Finally, it offers a systems-based analytical tool for organizing actions to promote the transition from environmental to ecological law.

Increasing the visibility, clarity and development of ecological law, this book will be of great interest to students and scholars of ecological and environmental law and governance.

Geoffrey Garver teaches part time at McGill University and Concordia University in Montreal, Canada, and coordinates research on law and governance for the Leadership for the Ecozoic program of McGill University and the University of Vermont. He is a member of the Steering Committee of the Ecological Law and Governance Association and is active in the international degrowth movement. He received his PhD (Geography) from McGill University in 2016 and has a B.S. (Chemical Engineering) from Cornell University, a J.D. from Michigan Law School and an LL.M. from McGill University.

Routledge Explorations in Environmental Studies

Romantic Anti-capitalism and Nature
The Enchanted Garden
Robert Sayre and Michael Löwy

Liberty and the Ecological Crisis
Freedom on a Finite Planet
Edited by Christopher J. Orr, Kaitlin Kish and Bruce Jennings

Environmental Justice and Oil Pollution Laws
Comparing Enforcement in the United States and Nigeria
Eloamaka Carol Okonkwo

The Environmental Impact of Overpopulation
The Ethics of Procreation
Trevor Hedberg

Riverlands of the Anthropocene
Walking Our Waterways as Places of Becoming
Margaret Somerville

An Environmental History of Australian Rainforests until 1939
Fire, Rain, Settlers and Conservation
Warwick Frost

Daoism and Environmental Philosophy
Nourishing Life
Eric S. Nelson

Ecological Law and the Planetary Crisis
A Legal Guide for Harmony on Earth
Geoffrey Garver

www.routledge.com/Routledge-Explorations-in-Environmental-Studies/book-series/REES

Ecological Law and the Planetary Crisis

A Legal Guide for Harmony on Earth

Geoffrey Garver

Routledge
Taylor & Francis Group
LONDON AND NEW YORK

earthscan
from Routledge

First published 2021
by Routledge
2 Park Square, Milton Park, Abingdon, Oxon OX14 4RN

and by Routledge
605 Third Avenue, New York, NY 10017

First issued in paperback 2022

Routledge is an imprint of the Taylor & Francis Group, an informa business

Publisher's Note
The publisher has gone to great lengths to ensure the quality of this reprint but points out that some imperfections in the original copies may be apparent.

British Library Cataloguing-in-Publication Data
A catalogue record for this book is available from the British Library

Library of Congress Cataloging-in-Publication Data
Names: Garver, Geoffrey, author.
Title: Ecological law and the planetary crisis : a legal guide for harmony on earth / Geoffrey Garver.
Description: Milton Park, Abingdon, Oxon ; New York, NY : Routledge, 2021. | Includes bibliographical references and index.
Identifiers: LCCN 2020023199 (print) | LCCN 2020023200 (ebook) | ISBN 9780367894511 (hardback) | ISBN 9781003019251 (ebook)
Subjects: LCSH: Environmental law. | Sustainable development–Law and legislation. | Environmental responsibility. | Human ecology.
Classification: LCC K3585 .G37 2021 (print) | LCC K3585 (ebook) | DDC 344.04/6–dc23
LC record available at https://lccn.loc.gov/2020023199
LC ebook record available at https://lccn.loc.gov/2020023200

ISBN 13: 978-0-367-61295-5 (pbk)
ISBN 13: 978-0-367-89451-1 (hbk)
ISBN 13: 978-1-00-301925-1 (ebk)

DOI: 10.4324/9781003019251

Typeset in Times New Roman
by Wearset Ltd, Boldon, Tyne and Wear

In fond memory of my parents, Anneliese and Newton Garver

Contents

List of tables xi
Acknowledgments xii
List of abbreviations xiv

Introduction 1

PART I
The role of law in the rise of the global ecological crisis 9

1 The Anthropocene as a frame of reference 11

2 The human dilemma in the Anthropocene 22

3 Energy transitions in history and their impacts 28

4 How conceptual factors like law helped lead to the Anthropocene 42

PART II
The failure of environmental law to respect key ecological limits 63

5 An overview of the failures of environmental law 65

6 Key examples of the inadequacy of environmental law 73

PART III
Ecological law: systems-based rules for a flourishing Earth 91

7 What is ecological law? 93

8 A mutually enhancing human-Earth relationship as the primary goal for law 97

9 The systems-based perspective underlying ecological law 108

10 The limits-insistent narrative and planetary boundaries 114

11 Eleven core features of ecological law 127

12 Ecological integrity and attachment to place in ecological law 152

PART IV
Getting from environmental to ecological law 167

13 A systems-based guide for moving from environmental to ecological law 169

14 Overcoming ecological challenges of international trade: an illustration 183

PART V
Degrowth and ecological law 225

15 The degrowth movement as a testing ground for ecological law 227

Index 235

Tables

9.1 Structure and key elements of legal systems 111
13.1 Leverage points in legal systems 172
13.2 Lock-in assessment: obstacles to coherent climate change law and policy 178
13.3 Lock-in assessment: opportunities for coherent climate change law and policy 179
14.1 Comparison of SEM process in NAAEC Articles 14 and 15 and Investor-State Dispute mechanisms in NAFTA Chapter 11 202
14.2 Leverage points: current approach compared to proposed approach for ensuring that trade and investment respects ecological limits 210
14.3 Summary of key lock-in tendencies that hinder international trade and investment that promotes a mutually enhancing human-Earth relationship 212
14.4 Lock-in assessment: obstacles in US law to international trade and investment that would promote and preserve a mutually enhancing human-Earth relationship 213
14.5 Lock-in assessment: opportunities for US law on international trade and investment that fosters and preserves a mutually enhancing human-Earth relationship 215

Acknowledgments

This book combines and updates my LLM thesis, which I completed in 2011, and my PhD thesis, which I completed in 2016. I am grateful for the numerous consultations, review comments and helpful guidance I received from my PhD thesis supervisor, Professor Peter G. Brown, and from my other thesis committee members, Professor Robert Godin (McGill Faculty of Law) and Professor Gail Chmura (McGill Department of Geography). I also acknowledge the guidance I received from prior members of my thesis committee, Professor Navin Ramakutty (formerly of the McGill Geography Department and now at the University of British Columbia) and Professor Daviken Studnicki-Gizbert (McGill Department of History). I am thankful as well for the numerous consultations, review comments and helpful guidance I received from my LLM thesis supervisor, Professor Richard A. Janda (McGill Faculty of Law).

Special thanks are due to Carla Sbert, Keith Helmuth, Judy Lumb, Pamela Haines and Dan Tarlock, who provided helpful comments on drafts of my LLM and PhD theses. I also benefited from comments I received on presentations of aspects of my research at numerous annual conferences of the Global Ecological Integrity Group, with particular thanks to Laura Westra and Klaus Bosselmann. I am grateful as well for the guidance, support and encouragement I received from McGill Professors Mark Goldberg, Jim Fyles and Nicolas Kosoy; members of the Steering Committee of the Economics for the Anthropocene (E4A) Partnership (Professors Jon Erickson, Josh Farley and Taylor Ricketts of the University of Vermont; and Professors Peter Victor and Ellie Perkins from York University); and the terrific students and staff in the E4A project. My fellow Steering Committee members of the Ecological Law and Governance Association, Klaus Bosselmann, Peter Burdon, Cristiane Derani, Katy Gwiazdon, Michelle Maloney, Lisa Mead, Massimiliano Montini and Ngozi Finette Unuigbe have greatly helped me develop my thinking about ecological law. I also thank my family, especially my late parents Newton and Anneliese Garver, for their steady support and encouragement as I developed the ideas presented in this book.

I am grateful for the financial support I received from McGill University, especially the Geography Department, for funding that Professor Peter G. Brown arranged for me through various Social Sciences and Humanities Research

Council (SSHRC) research grants and other sources, and for financial support I received from Hydro Québec for my LLM thesis.

This book incorporates revised and updated portions of the following published works. Every effort has been made to contact copyright holders for their permission to reprint material in this book. The publishers would be grateful to hear from any copyright holder who is not here acknowledged and will undertake to rectify any errors or omissions in future editions of this book.

Garver, G. 2012. Introducing the Rule of Ecological Law, book chapter in L. Westra, C. Soskolne and D. Spady, eds., *Human Health and Ecological Integrity*. New York: Earthscan.

Garver, G. 2013. The rule of ecological law: the legal complement to degrowth economics. *Sustainability* 5(1): 316–337.

Garver, G. 2013. Moving Forward with Planetary Boundaries and Degrowth, book chapter in L. Westra, A. Michelot and P. Taylor, eds., *Confronting Ecological and Economic Collapse*. New York: Earthscan.

Garver, G. 2015. A complex adaptive legal system for the challenges of the Anthropocene, Chapter 19 in L. Westra et al., eds., *Ecological Systems Integrity: Governance, Law and Human Rights*. New York: Earthscan.

Garver, G. 2015. Forgotten Promises: Neglected Environmental Provisions of the NAFTA and the North American Agreement on Environmental Cooperation, book chapter in H. Kong and K. Wroth, eds., *NAFTA and Sustainable Development*. New York: Cambridge University Press.

Garver, G., and M.S. Goldberg. 2015. Boundaries and Indicators: Conceptualizing and Measuring Progress Toward an Economy of Right Relationship Constrained by Global Ecological Limits, book chapter in P.G. Brown and P. Timmerman, eds., *Ecological Economics for the Anthropocene*. New York: Columbia University Press.

Garver, G. 2017. Ecological Integrity in the Anthropocene: Lessons for Law from Ecological Restoration and Beyond, book chapter in F-T. Gottwald, J. Gray and L. Westra, eds., *The Role of Integrity in the Governance of the Commons*. New York: Springer.

Garver, G. 2018. Moving From Environmental Law to Ecological Law: Frameworks, Priorities and Strategies, Chapter 14 in L. Westra, K. Bosselmann, J. Gray and K. Gwiazdon, eds., *Ecological Integrity, Law and Governance*. New York: Routledge.

Garver, G. 2019. A systems-based tool for transitioning to law for a mutually enhancing human-Earth relationship. *Ecological Economics* 157: 165–174.

Garver, G. 2019. Confronting remote ownership problems with ecological law. *Vermont Law Review* 43(3): 425–454.

Abbreviations

BII	Biological Integrity Index
CEC	Commission for Environmental Cooperation (North America)
CPR	Common pool resources
DR-CAFTA	Dominican Republic-Central America-US Free Trade Agreement
E4A	Economics for the Anthropocene
EIA	Environmental Impact Assessment
EPA	Environmental Protection Agency (United States)
G20	Group of Twenty nations
GAP	Group Assembly Process
GATS	General Agreement on Trade in Services
GATT	General Agreement on Tariffs and Trade
GDP	Gross domestic product
GHG	Greenhouse gas
GNP	Gross national product
GtC	Gigatonnes of carbon
HANPP	Human appropriation of net primary production
HDI	Human Development Index
IDS	Investor Dispute Settlement
IPAT	Impact is a function of Population, Affluence and Technology
IPCC	Intergovernmental Panel on Climate Change
JPAC	Joint Public Advisory Committee
MEFA	Material and energy flow accounting
MSY	Millions of species per year
NAAEC	North American Agreement on Environmental Cooperation
NAAQS	National Ambient Air Quality Standard (United States)
NAFTA	North American Free Trade Agreement
NGO	Nongovernmental organization
NPP	Net primary production
PB	Planetary boundary
SARA	Species at Risk Act
SDGs	Sustainable Development Goals
SEM	Submissions on Enforcement Matters
SPP	Security and Prosperity Partnership

TEIA	Transboundary Environmental Impact Assessment
TEU	Treaty on European Union
TFEU	Treaty on the Functioning of the European Union
TRIPS	Trade-Related Aspects of Intellectual Property Rights
TTIP	Transatlantic Trade and Investment Partnership
UNEP	United Nations Environment Programme
WTO	World Trade Organization

Introduction

Humanity faces critical choices as how collectively to approach the future in light of the alarming state of human-Earth relations. Humans have become a determinant, and increasingly destructive, factor in Earth's life systems. The ways in which humans have pursued and continue to pursue their collective endeavors is overwhelming Earth's capacity to support humanity in the long term. With a focus on law, this book uses a transdisciplinary systems perspective to analyze the role of material and conceptual factors that have led to this human-Earth dilemma and to propose a radical pathway to change direction toward a mutually enhancing human-Earth relationship.

Why law, and why a systems-based approach? In light of humanity's ecological dilemma, law stands out because it encodes many of the most hard-wired aspects of the ecologically destructive behaviors in the world's dominant societies and the global networks of power, governance, trade and information exchange that they have assembled and largely control. Law converts moral leanings and suggestions into enforceable requirements, prohibitions, freedoms and rights, and so laws have particular force ecologically. A systems approach reveals the ecological role of law. Legal systems are intertwined with social, political, cultural and ecological systems that continually adapt as they move together in time. In turn, law itself is dynamic and always evolving, despite the consistency, predictability and order that law generally is meant to instill. Several scholars theorize that legal systems are complex adaptive systems that evolve according to the same fundamental systems dynamics that underlie the evolution of other complex adaptive systems (Hornstein 2005; Ruhl 2008; Craig 2013; Garmestani and Benson 2013; Kim and Bosselmann 2013; Green et al. 2015). According to this theory, legal systems co-evolve with other complex adaptive social and ecological systems—collectively, an evolving web of systems called a "panarchy" (Holling et al. 2002)—through interactions and feedbacks at and across multiple temporal and spatial scales (Brooks et al. 2002). This co-evolution involves a constant "interplay between change and persistence, between the predictable and unpredictable" (Holling et al. 2002: 1).

How legal systems interact and evolve with other subsystems of the human-Earth system takes on heightened importance in the context of a preliminary determination that, as of around 1950, Earth has entered a new, human-dominated period called the Anthropocene (Anthropocene Working Group 2019). The Earth's life systems, and humanity itself, face a predicament without historical precedent due to expanding aggregated impacts of human activity that by several measures have already overshot global ecological limits of a world fit for humans and many other living beings and ecosystems (Rockström et al. 2009; Carpenter and Bennett 2011; Steffen et al. 2011; Steffen et al. 2015). In the past two decades, many scientists have been calling for formal geological recognition of the Anthropocene—a period of time most often associated with the onset of the Industrial Revolution in the late eighteenth century (Crutzen and Stoermer 2000; Rockström et al. 2009; Steffen et al. 2011; Steffen et al. 2015). Although not yet formally recognized as a geological epoch, the Anthropocene underscores the profound temporal and spatial reach of humanity's ecological impacts and puts in perspective a new way to consider the past, present and future of the human-Earth relationship.

Building on the notion of legal systems as complex adaptive systems within the global panarchy, along with other recent proposals on how better to align legal systems with the life-enhancing functioning of Earth systems, this book develops a framework for law in the Anthropocene. A systems-based approach to legal systems and the other social and ecological systems with which law co-evolves across temporal and spatial scales is central to this analysis and the proposals and examples that emerge from it. The normative core of this approach is the notion of a mutually enhancing human-Earth relationship (Berry 1999). In a mutually enhancing human-Earth relationship, humans individually and collectively see themselves as members, not masters, of the entire community of life on Earth and interact with Earth and the life it supports respectfully and "for the benefit of the larger community as well as ourselves" (Berry 1999: 5). The empirical foundation of this approach derives from the historical evolution of, and future prospects for, the stocks and flows of matter and energy that support the human enterprise, and the ecological impacts of this "social metabolism" (Fischer-Kowalski and Amann 2001: 12). I and others call the law that is grounded in a mutually enhancing human-Earth relationship and that justly and fairly regulates humanity's social metabolism within ecological limits, *ecological law*: a new kind of law that responds to the current planetary crisis and supports harmony on Earth.

In Part I of this book, I examine the role of law in the rise of the global ecological crisis. Part I begins with an analysis of the concept of the Anthropocene and how its emergence correlates with a growing dilemma in the human-Earth relationship. I then review key material and conceptual factors that have led to this dilemma, with an emphasis on the role of law as a significant conceptual factor. The key material factors derive from the shifts in the energetics of human activities that occurred with the Neolithic Transition around 10,000 years ago (from hunting and gathering societies to agriculture societies) and the Industrial Revolution that began in the late eighteenth century (from agricultural societies mostly powered by sun, wind, water, plants, animals and people to industrial societies mostly powered by fossil fuels). Both these shifts had significant impacts on the distribution and growth of the human population and on the human-Earth relationship. The unrivaled human capacity for collective learning, developed well before the rise of agricultural societies, underlies all conceptual factors that have led human society to the present circumstances—and also underlies prospects for humanity collectively to transition to a better human-Earth relationship in the future (Rees 2002; Harari 2014).

Law and other collectively constructed normative institutions emerged as enduring features of social and political systems with the shift from hunting and gathering to agriculture and the consequent rise of permanent settlements and eventually more complex civilizations. Denser human settlements led to persistent social hierarchies, ever-increasing specialization of roles in society, strong notions of property-based entitlements, inequalities in wealth, increasing dependence on remote places for provisioning of human needs and wants and a tendency toward expansion and development. Modern legal systems from the local to the global level have roots in that transition, as do the interconnected social, economic and political systems to which law applies (Burdon 2015). A dominant international commitment to perpetual economic growth, which the law sustains through enforceable rules, has become the central conceptual driver of the most critical challenges of the human-Earth relationship in the Anthropocene (McNeill 2000; Speth 2008; Steffen et al. 2011; United Nations 2012; United Nations 2015a; G20 2019). Part I of this book calls sharply into question this commitment to infinite economic growth on a finite planet.

Part II explains how the regimes of environmental law that arose in the latter part of the twentieth century have badly failed to maintain humanity's social metabolism within crucial ecological limits, from the local to the global scale. For all its success in limiting pollution and revealing environmental impacts of human activities, environmental law has remained firmly tethered to legal, economic and political systems with an overarching goal of maintaining perpetual economic growth. In these systems, environmental protection ultimately gives way to economic, political and technological considerations, despite long-term adverse consequences. By contrast, ecological law is grounded in the notion that economic and socio-political spheres must be subservient to global ecological limits beyond which lie catastrophic ecological, social and economic collapse. Planetary boundaries of safe operating space for humanity, along with complementary measures and principles related to both ecological limits and fair sharing, provide the scientific and ethical basis of ecological law. I illustrate the limitations of contemporary environmental law with mostly ineffective attempts to address systemic environmental problems like eutrophication, acid rain, climate change and depletion of the ozone layer, followed by a critique of the broader legal and institutional architecture now in place to address global environmental challenges. The current approach to international trade and the environment highlights the failure of the global governance infrastructure to respect global ecological boundaries on the human enterprise.

Part III proposes orienting legal systems toward a more hopeful trajectory for humans and Earth's life systems, grounded in science, systems thinking and intergenerational and interspecies fairness. I explain why fostering and maintaining a mutually enhancing human-Earth relationship (Berry 1999) within a limits-insistent systems framework is a more desirable overarching objective for humanity, and hence for law, than the current dominant commitment to perpetual economic growth. Because this objective seeks flourishing of both humans and the diverse ecosystems and species that maintain humanity, the orientation I propose is ecocentric, with the explicit condition that harmonious inclusion of humans is essential—something more radical versions of ecocentrism might omit. Part III provides an analysis in depth of this more desirable societal goal, how it differs from the growth-insistent normative regime and how it provides a foundation for a more enduring long-term human-Earth relationship. I present 11 features of ecological law (Garver 2013) that derive from the overarching quest for a mutually enhancing human-Earth relationship:

1. Humans are part of the Earth's life systems, not separate from them, and concepts that are relevant to ecological law, like ecological integrity, must include humans within flourishing life systems;
2. Ecological limits, such as planetary boundaries for safe operating space for humanity, must have primacy over economic, political and other considerations;
3. Ecological law permeates the entire legal regime in a systemic, integrated way, and is not a specialty area of the law like environmental law;
4. In light of the current ecological crisis, ecological law should focus on reducing the throughput of material and energy in the human economy;
5. Ecological law must ensure that biomass and extracted materials are obtained and used in the economy according to real needs, with minimal consideration of utilitarian desires reflected in market prices;
6. Ecological law must be global but applied using principles of proportionality (sufficient but not excessive regulation) and subsidiarity (application of law at the lowest political tier at which the law's objective can be achieved);
7. Ecological law must ensure fair sharing of resources among present and future generations of humans and other life;
8. Ecological law must be binding and supranational, with supremacy over sub-global legal regimes as necessary;
9. A greatly expanded program of research and monitoring tied to improved understanding and continual adjustment of ecological boundaries and means for respecting them is needed to support ecological law from the global to the local level;
10. Ecological law requires precaution about crossing planetary boundaries, with both margins of safety to ensure that the boundaries are respected and complementary measures, such as ecological regeneration and restoration, to allow the Earth's life systems to flourish;
11. Ecological law must be adaptive, in recognition of the non-equilibrium nature of ecosystems and the need to get started on a comprehensive effort to constrain the economy within ecological limits despite uncertainty.

These 11 features indicate that even if some ecological law must operate at a global level, as with climate change, overall, it will favor place-based approaches that meet certain conditions. The work of Ostrom (1990) and others on how various communities have developed place-based rules and approaches for using water, timber and other resources sustainably in perpetuity provides useful insights for how the 11 features of ecological law might arise and be implemented. Principles and approaches used in the applied science of ecological and eco-cultural restoration provide additional guidance for giving coherence to relevant concepts such as ecological integrity, which often is aligned with pristine nature unaffected by humans, so that those concepts can better be applied from the local to the global scale.

In Part IV, I present a systems-based guide for moving from contemporary law, including environmental law, to ecological law. First, I explain how a systems-based perspective of law as a complex adaptive system that interacts and evolves along with other systems in a global panarchy (Gunderson and Holling 2002) can provide the foundation for a transition to a mutually enhancing human-Earth relationship in the Anthropocene. Because legal systems behave as complex adaptive systems that interact with other such systems, an adaptive approach to law that is more responsive to and coherent with processes of global environmental change and their effects on the human prospect is needed. In adapting law to change, transformation of legal systems per se is less important than ongoing adaptation of how law mediates humanity's relationship with the finite life support capacity of Earth systems.

To guide this adaptive transformation from contemporary law to ecological law, I have developed a lock-in/lock-out assessment tool that uses systems leverage points and assessments of resistance to desired change. Although I present this tool in a particular form, the underlying concepts are applicable to a broad plurality of situations and context, and it is intended to be flexible and adaptable. Systems-based leverage points (Meadows 1999) for triggering change in systems structure and function can be either superficial or deep-seated, and they are useful for developing strategies for adapting legal systems, or systems with which law interacts, in desired ways. Along with leverage points, assessment of the degree to which desirable or undesirable features are locked in or locked out of legal systems or other systems with which legal systems interact are the second element of my proposed tool for revealing these strategies. For example, constitutions are legal instruments that are relatively hard to amend, and therefore they have a relatively high degree of lock-in. In assessing the degree of lock-in within a system, the roles of path dependence, resilience and adaptiveness—all core elements of complex adaptive systems—are particularly important.

The key metrics and indicators that support this assessment methodology, introduced briefly in Part II, pertain to the impacts of human manipulation of material and energy stocks and flows on Earth's life support systems. These metrics and indicators include the planetary boundaries concept and framework (Rockström et al. 2009; Steffen et al. 2015), the IPAT framework[1] (Ehrlich and

Holdren 1972) and others, such as ecological footprint and human appropriation of net primary production (HANPP).[2] They can be applied in assessments of leverage points of change and degrees of lock-in or lock-out that resists change to yield strategies and priorities for moving toward and preserving a mutually enhancing human-Earth relationship across spatial and temporal scales.

After describing generally how lock-in/lock-out assessments work in Part IV, I give an example of how they can be used to develop strategies for overcoming the enormous ecological challenges related to the current system of international trade, which many deeply entrenched laws, policies and institutional structures sustain. This analysis reveals that the approach to international trade and finance reflected in the World Trade Organization (WTO) and most bilateral and multilateral trade and investment agreements increases the degree of lock-in of system features that impede progress toward a mutually enhancing human-Earth relationship. In particular, analysis of the first 20 years of North American Free Trade Agreement (NAFTA) shows its environmental provisions, along with the environmental side agreement to NAFTA, present a near ideal example of a tragic institution: that is, an institution that addresses aspects of the Tragedy of the Commons that has become ineffective or worse but difficult to change (Daniels 2007). NAFTA's replacement, the United States-Mexico-Canada Agreement (USMCA), contains little or nothing to change that description. I propose an alternative, human-inclusive ecocentric approach to international trade, built on the goals, concepts and metrics for ecological law developed in the preceding chapters. Such a trade regime would require a global framework founded on an international commitment to a mutually enhancing human-Earth relationship, ecological integrity and fair sharing of Earth's life support capacity among present and future generations of humans and the rest of Earth's commonwealth of life (Brown and Garver 2009).

In Part V, I propose a policy-oriented approach combining research and activism for identifying and pursuing strategies and priorities to achieve a mutually enhancing human-Earth relationship in a regime of ecological law. The proposed agenda resonates in many ways with the research and activism approach of the nascent but building *décroissance* (degrowth) movement that has come to life since 2000 (Research & Degrowth 2016). Degrowth "refers to a trajectory where the 'throughput' (energy, materials and waste flows) of an economy decreases while welfare, or well-being, increases" (Kallis 2018: 9). The degrowth movement is therefore in many ways an ideal testing ground for ecological law. My proposed approach also fits well with a wide range of other social movements seeking transformative improvement in human-Earth relations, such as the Transition Town movement and communities orienting toward the concept of *buen vivir*. The increasingly fragile prospects for achieving a mutually enhancing human-Earth relationship in the Anthropocene in light of trends in the human-Earth system underscore the urgent need to undertake rigorous lock-in/lock-out assessments in order to identify strategies and priorities across scales and in a broad plurality of contexts and situations.

Notes

1 The IPAT framework is used to assess how population (P), affluence (or, more appropriately, consumption) (A) and technology (T) interact to produce an overall impact (I), such as climate change or biodiversity loss.
2 HANPP is a measure of how much of the net primary production produced through photosynthesis is appropriated by humans, either directly through harvest or indirectly through, for example, urbanization or other land use changes.

References

Anthropocene Working Group. 2019. Results of binding vote by AWG, release 21st May 2019. Subcommission on Quaternary Stratigraphy [online]: http://quaternary.stratigraphy.org/working-groups/anthropocene/

Berry, T. 1999. *The Great Work: Our Way Into the Future*. New York: Three Rivers Press.

Brooks, R.O., R. Jones and R.A. Virginia. 2002. *Law and Ecology: The Rise of the Ecosystem Regime.* Aldershot UK: Ashgate Publishing Ltd.

Brown, P.G., and G. Garver. 2009. *Right Relationship: Building a Whole Earth Economy.* San Francisco: Berrett-Koehler.

Burdon, P.D. 2015. *Earth Jurisprudence: Private Property and the Environment.* New York: Routledge.

Carpenter, S.R., and E.M. Bennett. 2011. Reconsideration of the planetary boundary for phosphorus. *Environmental Research Letters* 6: 1–12.

Craig, R.K. 2013. Learning to think about complex environmental systems in environmental and natural resources law and legal scholarship: a twenty-year retrospective. *Fordham Environmental Law Review* 24: 87–102.

Crutzen, P.J., and E.F. Stoermer. 2000. The Anthropocene. *Global Change Newsletter* 41: 17–18.

Daniels, B. 2007. Emerging commons and tragic institutions. *Environmental Law* 37(3): 515–571.

Ehrlich, P.R., and J.P. Holdren. 1972 (May). Critique of *The Closing Circle*. *Bulletin of the Atomic Scientists*, 16–27.

Fischer-Kowalski, M., and C. Amann. 2001. Beyond IPAT and Kuznets Curves: globalization as a vital factor in analysing the environmental impact of socio-economic metabolism. *Population and Environment* 23(1): 7–47.

G20. 2019. *G20 Osaka leaders' declaration.* [online]: www.g20.utoronto.ca/2019/2019-g20-osaka-leaders-declaration.html.

Garmestani, A.S., and M.H. Benson. 2013. A framework for resilience-based governance of social-ecological systems. *Ecology and Society* 18(1) art 9 [online]: www.ecologyandsociety.org/vol.18/iss1/art9/.

Garver, G. 2013. The rule of ecological law: the legal complement to degrowth economics. *Sustainability* 5(1): 316–337.

Green, O.O., A.S. Garmestani, C.R. Allen, L.H. Gunderson, J.B Ruhl, C.A. Arnold, N.A.J. Graham, B. Cosens, D.G. Angeler, B.C. Chaffin and C.S. Holling. 2015. Barriers and bridges to the integration of social-ecological resilience and law. *Frontiers in Ecology and the Environment* 13(6): 332–337.

Gunderson, L.H., and C.S. Holling, eds. 2002. *Panarchy: Understanding Transformations in Human and Natural Systems.* Washington, DC: Island Press.

Harari, Y.N. 2014. *Sapiens: A Brief History of Humankind.* Toronto: McClelland & Stewart.

Holling, C.S., L.H. Gunderson and D. Ludwig. 2002. In Quest of a Theory of Adaptive Change, pp. 3–22 in L.H. Gunderson and C.S. Holling, eds., *Panarchy: Understanding Transformations in Human and Natural Systems.* Washington, DC: Island Press.

Hornstein, D.T. 2005. Complexity theory, adaptation, and administrative law. *Duke Law Journal* 54: 913–960.

Kallis, G. 2018. *Degrowth.* Newcastle upon Tyne, UK: Agenda Publishing.

Kim, R.E., and K. Bosselmann. 2013. International environmental law in the Anthropocene: towards a purposive system of multilateral environmental agreements. *Transnational Environmental Law* 2: 285–309.

McNeill, J.R. 2000. *Something New Under the Sun: An Environmental History of the Twentieth-Century World.* New York: W.W. Norton & Co.

Meadows, D. 1999. *Leverage Points: Places to Intervene in a System.* Hartland, VT: The Sustainability Institute.

Ostrom, E. 1990. *Governing the Commons: The Evolution of Institutions for Collective Action.* New York: Cambridge University Press.

Rees, W.E. 2002. Globalization and sustainability: conflict or convergence? *Bulletin of Science, Technology & Society* 22.

Research & Degrowth. 2016. *Research and actions to consume less and share more* [online]: www.degrowth.org.

Rockström, J., W. Steffen, K. Noone, Å. Persson, F.S. Chapin, III, E. Lambin, T.M. Lenton, M. Scheffer, C. Folke, H. Schellnhuber, B. Nykvist, C.A. De Wit, T. Hughes, S. van der Leeuw, H. Rodhe, S. Sörlin, P.K. Snyder, R. Costanza, U. Svedin, M. Falkenmark, L. Karlberg, R.W. Corell, V.J. Fabry, J. Hansen, B. Walker, D. Liverman, K. Richardson, P. Crutzen and J. Foley. 2009. Planetary boundaries: exploring the safe operating space for humanity. *Ecology and Society* 14(2): 32. [online]: www.ecologyandsociety.org/vol.14/iss2/art32/

Ruhl, J.B. 2008. Law's complexity: a primer. *Georgia State University Law Review* 24(4): 885–911.

Speth, J.G. 2008. *The Bridge at the Edge of the World.* New Haven, CT: Yale University Press.

Steffen, W., J. Grinevald, P. Crutzen and J. McNeill. 2011. The Anthropocene: conceptual and historical perspectives. *Philosophical Transactions of the Royal Society A* 369: 842–867.

Steffen, W., K. Richardson, J. Rockström, S.E. Cornell, I. Fetzer, E.M. Bennett, R. Biggs, S.R. Carpenter, W. de Vries, C.A. de Wit, C. Folke, D. Gerten, J. Heinke, G.M. Mace, L.M. Persson, V. Ramanathan, B. Reyers and S. Sörlin. 2015. Planetary boundaries: guiding human development on a changing planet. *Science* 347(6223): 1259885-1–1259885-10.

United Nations. 2012. *The future we want.* New York: United Nations [online]: https://documents-dds-ny.un.org/doc/UNDOC/GEN/N11/476/10/PDF/N1147610.pdf?OpenElement.

United Nations. 2015a. *Transforming Our World: The 2030 Agenda for Sustainable Development.* New York: United Nations.

Part I

The role of law in the rise of the global ecological crisis

1 The Anthropocene as a frame of reference

In Part I, I present a diagnosis of the current human-Earth dilemma and how law and legal systems are implicated in it. The concept of a new geological epoch called the "Anthropocene" provides the overall context within which to analyze this dilemma. Preliminarily recognized in May 2019 as a new geological epoch by the Anthropocene Working Group (AWG) of the Subcommission on Quaternary Stratigraphy of the International Commission on Stratigraphy (Anthropocene Working Group 2019), the Anthropocene puts in perspective a new way to examine the past, present and future of the human-Earth relationship. It forces attention on humans as an integral and increasingly influential element of the global Earth system and the complex interconnected subsystems that it comprises. I start the diagnosis in Part I by proposing that, though flawed, the Anthropocene concept is a useful frame of reference for examining the past, present and future role of law in the human-Earth relationship and for understanding the dilemma in human-Earth relationship that has emerged along with the Anthropocene. I turn then to an analysis of key human-driven material factors that have contributed to the current dilemma and the emergence of the Anthropocene. Finally, I consider and analyze how law and legal systems and related conceptual factors contributed to the rise of the global ecological crisis.

A clash of mutually exclusive narratives looms in the background of humanity's ecological dilemma. The growing tension between narratives based on either insistence on economic growth or insistence on an economy contained within ecological limits creates an opportunity for transformation of socially constructed norms away from the growth-insistent narrative. Legal systems are one such construct. An examination of how law has contributed to the onset of the Anthropocene and the current globally dominant growth-insistent narrative puts into perspective prospects for transforming law and legal systems to help humanity foster a mutually enhancing human-Earth relationship. This examination includes consideration of how law and legal systems interact with other normative and ecological systems that make up the panarchy (Gunderson and Holling 2002) in which global change occurs. This panarchy encompasses the entire integrated complex of evolving economic, ecological, social and institutional systems that interact across temporal and spatial scales, accounting for rapid, patent changes as well as slow changes that are often unpredictable or difficult to perceive, and for the chaos, emergence and catastrophe that are axiomatic features of complex dynamic systems (Holling et al. 2002; Ruhl 2012). Thus, the diagnosis in Part I is grounded in a transdisciplinary systems view of global change, a perspective that necessarily accounts for complexity and irreducible uncertainty in the evolving human-Earth system.

The concept of the Anthropocene underscores the notion that because of its entrenched and irreversible ecological impacts over thousands of years, humanity has reached a point of no return. The Anthropocene cannot be undone (Ellis 2011); it can only frame the current human-Earth dilemma and potential responses to it, including those that will improve the human prospect within Earth's limited life support capacity. In geology, stratigraphic markers of epochs are determined according to their scale, magnitude and significance, and to be valid must be more or less permanent and of the same magnitude as markers of earlier epoch boundaries (Zalasiewicz et al. 2011a; Waters et al. 2014, 2016). Unlike other formally recognized geological time periods, the Anthropocene represents both a point of no return and a glaring spotlight on serious dangers in the human-Earth relationship. For the reasons presented below, it is a useful and compelling concept for analyzing and framing responses to the current human-Earth dilemma.

Whether humanity's impacts have left the kind of unique, enduring marker or set of markers required for formal recognition of a new geological epoch is for geologists to determine (Zalasiewicz et al. 2011a; Waters et al. 2014; 2016). Examples include changes in fossil assemblages following an extinction event or a significant change in biodiversity, sharp changes in the chemical composition of sediments or the oceans due to alteration of geochemical cycles and modification of sedimentation patterns (Zalasiewicz et al. 2011a). A recent proposal to consider "technostratigraphical" markers that combine the geological impacts of many different forms of human technology is another potential candidate (Zalasiewicz et al. 2017; Waters et al. 2016). In 2009, the AWG undertook a formal process for considering the validity of the Anthropocene as a geological epoch (Corlett 2015). In light of considerable resistance to formally recognizing the Anthropocene at all (Lewis and Maslin 2015; Gibbard and Walker 2014; Visconti 2014; Waters et al. 2016), as well as differing proposals for the start of the Anthropocene, the outcome was far from certain (Castree 2014; Corlett 2015). One likely alternative to formally recognizing the Anthropocene was "to leave it as an informal, albeit solidly founded, geological time term, as the Pre-Cambrian and Tertiary currently are" (Waters et al. 2016: aad2622–8).

The most common assertion was that the Anthropocene began at the start of the Industrial Revolution in the late eighteenth century (Crutzen and Stoermer 2000; Rockström et al. 2009; Steffen et al. 2011; Zalasiewicz et al. 2011b; Oldfield et al. 2014). Humans have been driving extensive alteration of the ecosystems in which they are embedded since well before the Industrial Revolution. Some researchers asserted the Anthropocene began well before the industrial era with the extensive land use changes that emerged in the Neolithic Transition from hunter-gatherer societies to agricultural societies (Steffen et al. 2011;

Ruddiman 2013; Smith and Zeder 2013; Corlett 2015) or possibly even with human-driven extinction of megafauna in Paleolithic times (Steffen et al. 2011; Foley et al. 2013; Corlett 2015). More commonly, the term Anthropocene calls particular attention to the ecological impacts of the explosive expansion of human society associated with fossil fuels, and especially the Great Acceleration of human-driven global change that has been occurring since the middle of the twentieth century (Steffen et al. 2011; Zalasiewicz et al. 2011b; Oldfield et al. 2014; Waters et al. 2016).

In light of these different perspectives, the Anthropocene concept focused attention on three stages: pre-industrial (pre-1800), early industrial (1800–1950) and the Great Acceleration (post-1950) (Oldfield et al. 2014), with a fourth "stewardship" stage possibly emerging (Steffen et al. 2007; Castree 2014). This proposal of a stewardship stage resonates well with the analysis and proposals in this book. Alternatively, recognizing the long history leading to the current degree of human impact on Earth systems, some proponents of referring to the Anthropocene as the period since the Industrial Revolution would label the prior period of significant human impact as the Paleoanthropocene (Foley et al. 2013). Others noted that the Holocene is already distinguished from the Pleistocene in part because of the influence of humans, and therefore that the term Holocene is sufficient to include any subsequent human impacts of geological significance (Lewis and Maslin 2015).

Ultimately, the AWG determined that the Anthropocene should be formally recognized as a new epoch defined by a Global boundary Stratotype Section and Point (GSSP), or "golden spike," optimally in the middle of the twentieth century (Anthropocene Working Group 2019). However, the International Union of Geological Sciences has not yet formally recognized the Anthropocene in early 2020. The AWG is developing a proposal for formal recognition, with the mid-twentieth century golden spike "coinciding with the array of geological proxy signals preserved within recently accumulated strata and resulting from the 'Great Acceleration' of population growth, industrialization and globalization" (Anthropocene Working Group 2019). Radionuclides from thermonuclear bomb tests beginning in the 1950s are the clearest and most global marker from this time period.

Despite its geological resonance, the term "Anthropocene" emerged outside of geology (Crutzen and Stoermer 2000) as a way to highlight evidence that humanity "now rivals some of the great forces of Nature in its impacts on the functioning of the Earth system" (Steffen et al. 2011: 843). The broader scientific community is increasingly embracing the term Anthropocene, often without rigorous consideration of its geological implications. Nonetheless, the AWG recognizes that the Anthropocene is "a popular scientific term used by scientists, the scientifically engaged public and the media to designate the period of Earth's history during which humans have a decisive influence on the state, dynamics and future of the Earth System" (Anthropocene Working Group 2019).

For the social sciences and humanities, the geological inquiry is tangential; the Anthropocene has more to do with considering how ethical, social, cultural, political, economic and legal systems have contributed to dangerous accumulation of human impacts on Earth's life systems, and how human society might adjust those socially constructed systems to ward off risks of catastrophic evolution of Earth systems and promote an enduring and thriving human-Earth relationship (Kotchen and Young 2007; Kellie-Smith and Cox 2011; Zalasiewicz et al. 2011b; Robinson 2014; Oldfield et al. 2014; Malm and Hornborg 2014; Horn and Bergthaller 2020). Regardless of the geological validity of the Anthropocene, its material and conceptual drivers, "if they continue unabated through this century, may well threaten the viability of contemporary civilization and perhaps even the future existence of *Homo sapiens*" (Steffen et al. 2011: 862). This dilemma and the material and conceptual forces behind it are examined further in the next section.

The question remains whether the Anthropocene is a useful frame of reference for considering the past, present and future of the human-Earth relationship. An overarching critique is that valid stratigraphic markers of this new proposed epoch will not necessarily signal a problem or dilemma in the human-Earth relationship; they will merely establish a geological phenomenon. However, potential markers indicating the Anthropocene beginning sometime within the last 250 years correlate with scientific analyses indicating that global environmental change that would be catastrophic for the human-Earth relationship has begun or is likely without unprecedented adjustments in humanity's socially constructed normative systems (Speth 2008; Rockström et al. 2009; Carpenter and Bennett 2011; Steffen et al. 2011; McLellan et al. 2014; Steffen et al. 2015). In particular, the Anthropocene concept resonates strongly with the more explicitly normative planetary boundaries framework, which is largely keyed to maintaining Earth's social and ecological systems within the human-friendly range of the Holocene (Rockström et al. 2009; Castree 2014; Steffen et al. 2015). Indeed, the alignment of the potential factual basis in geology for the Anthropocene with concern over an unprecedented crisis in the human-Earth relationship is why the term was coined. The AWG recognized as much in its 2019 determination. Yet, the Anthropocene draws attention not only to human responsibility for anthropogenic drivers of momentous environmental change, but also to the human potential to respond to the dilemma that this unprecedented change has brought about. The implicit focus on the moral choice of what kind of Anthropocene humanity should pursue is part of what gives force and appeal to the Anthropocene concept (Ellis and Trachtenberg 2013).

In a prominent critique, Malm and Hornborg (2014) raise several other concerns that merit consideration. Accepting for purposes of their critique that the Anthropocene began with the Industrial Revolution and the rise of fossil fuel use, they fault the "Anthropocene narrative" on several grounds. First, they claim it is too beholden to the natural sciences. They also maintain that it is too much based on an implicit indictment of human nature and humankind generally. To the extent the Anthropocene implies problems with unique technological prowess of *Homo sapiens,* they find the causal connection between early humans' use of fire and more recent technologies such as the steam engine (which some defenders of the Anthropocene concept have noted as key to the emergence of the Anthropocene) to be too remote to be convincing. Instead, they find the Anthropocene narrative to be insufficiently cognizant of the role of unequal distribution of power and wealth—and in particular the role of a small elite that has historically controlled the means of production—in the development of the technologies and processes that are driving the global change that defines the Anthropocene.

Malm and Hornborg's (2014) objection to attributing to all humanity, and to human nature, the technologies and processes that have led to climate change and other anthropogenic ecological impacts has considerable merit. In particular, treating the Anthropocene as the inevitable result of human nature tends to minimize the prospects for human ingenuity and innovation—whether in technology or, more importantly, in the social, economic, legal, political and cultural spheres—to yield enduring solutions that avoid catastrophic environmental and social change. In addition to maintaining that a small minority of humans is historically responsible for the majority of climate change and other ecological disruptions, they further question any general correlation between those disturbances and humanity in general by presenting research showing that population (at least recently) tends to grow more slowly in the regions that are responsible for a disproportionately higher share of greenhouse gas emissions. On evidence such as this, they and others (Moore 2017) object to a species-encompassing term such as *Anthropo*cene.

Ultimately, their critique is unconvincing. First, it rests on an overstatement of what they call the Anthropocene narrative, which they mostly attribute to Crutzen (2002) and a few other natural scientists. Rather than being an "entrenched concept and mode of thinking" (Malm and Hornborg 2014: 5), the Anthropocene is still an emerging concept, and it is increasingly framing research that bridges the divide between the natural and social sciences (Castree 2014; Oldfield et al. 2014). Malm and Hornborg are certainly correct in asserting that the Anthropocene deserves much more attention in the social sciences and humanities (Ellis and Trachtenberg 2013), but this is happening to a greater extent than they acknowledge. Researchers have begun to investigate how innovations in law (Vidas 2011; Garver 2013; Kim and Bosselmann 2013; Scott 2013; Kotzé 2014; Robinson 2014), economics (Kellie-Smith and Cox 2011; Kosoy et al. 2012) ethics (Gibson-Graham and Roelvink 2009; Minteer 2012; Hourdequin 2013; Preston 2014) and other normative disciplines in the social sciences and humanities (Kotchen and Young 2007; Lövbrand et al. 2009; Slaughter 2012; Biermann 2014; Guillaume 2014, Lövbrand et al. 2015; Horn and Bergthaller 2020), as well as in interdisciplinary or transdisciplinary approaches that include the social sciences and humanities (Westley et al. 2011; Palsson et al. 2013; Fischer-Kowalski et al. 2014), might steer humanity away from the human-Earth dilemma in the Anthropocene (Oldfield et al. 2014). For example, the Economics for the Anthropocene partnership, launched in 2014 and later called Leadership for the Ecozoic, is strongly oriented toward regrounding normative disciplines such as economics, finance, law, governance and ethics so as to respond to the challenges for the human-Earth relationship in the Anthropocene (Economics for the Anthropocene 2016). As well, and contrary to the argument that the Anthropocene concept is entrenched and settled, the journal *Anthropocene Review* (which published Malm and Hornborg's critique) was established in 2014 to promote a transdisciplinary, systems-based approach to considering the evolving Anthropocene concept (Oldfield et al. 2014).

Moreover, even if Crutzen and others have noted causal roots of the Anthropocene in early humans' use of fire, they and many others use the term precisely to call attention to the need to reorient human societies away from the dilemma in the human-Earth relationship that aligns with potential geological markers of the Anthropocene. Thus, those researchers implicitly reject Malm and Hornborg's (2014) suggestion that the current situation is necessarily the inevitable result of deeply-rooted human traits that render it unsolvable. To the contrary, the Anthropocene concept helps drive a more urgent quest for "good" versions of the Anthropocene (Berkhout 2014) that could emerge as a result of various responses, both technological (at the extreme, the more radical geoengineering proposals (NRC 2015a; NRC 2015b))[1] and socio-political (such as using planetary boundaries to develop "novel and adaptive governance approaches at global, regional and local scales" (Rockström et al. 2009: 28)).

Further, Malm and Hornborg's (2014) concern that the natural sciences unduly dominate the Anthropocene concept does not recognize that whether or not it is formally recognized in geology, the Anthropocene concept has a strong normative bearing. For example, after analyzing several potential starting points for the Anthropocene, Lewis and Maslin (2015) note the differences in potential responses within the social sciences to the two candidates they find most convincing: (1) the drop in CO_2 in glacial ice that peaked in 1610 due to reforestation and land use changes resulting from sudden and extensive decreases in human population in the Americas following the first arrival of Europeans, and (2) the radionuclides in tree rings that peaked in 1964 as a result of the detonation of nuclear weapons. In the first case, they suggest that the relevant geological marker would indicate that "colonialism, global trade and coal brought about the Anthropocene" (Lewis and Maslin 2015: 177). In the second case, a marker connected to nuclear weapons would sharpen the focus on "a story of an elite-driven technological development that threatens planet-wide destruction" (Lewis and Maslin 2015: 177), as well as on humanity's collective capacity to ward off a global ecological and social threat through agreements to prohibit the use of nuclear weapons. Both cases provide foundations for normative responses to the human impacts associated with the Anthropocene.

Finally, as Malm and Hornborg (2014) acknowledge, even if a small subset of humanity is most directly responsible for climate change and other ecological markers of the Anthropocene, those impacts nonetheless arose within humanity writ large. Their argument depends on an artificial social and historical isolation of the owners of the means of production, who they see as the primary drivers of the fossil fuel-driven industrial processes in capitalist societies that have led to the current crisis in the human-Earth relationship. Putting aside questions of accountability and responsibility to which this observation is clearly relevant, their argument falls short. The small segment of humanity on which they focus is embedded in a social, cultural and historical context with a much broader sweep, as are the power relationships and socio-cultural dynamics that underlie this elite's societal position. Since the emergence of permanently settled agricultural societies, elites with disproportionate power and wealth have routinely held sway over increasingly stratified societies, living off agricultural surpluses produced (until workers started to transition from farms to factories in the industrial era) by the vast majority of people (Testart 1982; Davies 2002; Ponting 2007). Those elites, who have arisen independently in diverse socio-cultural contexts and in many kinds of political economies in addition to capitalism (McNeill 2000; Ponting 2007; Burke and Pomeranz 2009), could not exist outside of the broader social and ecological context that supports them. Malm and Hornborg (2014), along with others (Moore 2017), have good cause to focus particular attention on the capitalists that arose with sustained power alongside European colonialism in the early modern era. Those who prefer Capitalocene over Anthropocene have a strong point (Moore 2017). Nonetheless, the operations of the owners of production in the industrial age are thoroughly dependent on an increasingly globalized complex of social and economic infrastructure and systems implicating a wider swath of humanity than Malm and Hornborg acknowledge.

Instead of providing reasons for rejecting the Anthropocene concept, Malm and Hornborg (2014) provide sound reasons for resisting a fixed notion of it and for enhancing research on its implications in the social sciences. The Anthropocene highlights the stakes facing the entire community of life on Earth, but it poses a particular dilemma and opportunity for humanity. Whether the result of the inevitable evolution of human traits that date back to Paleolithic times, or of a random accident in the history of humans' intellectual capacity that allowed the transition to fossil-fuel powered industrial societies, the Anthropocene reflects the immense power of humans' unique—or at least unrivaled—capacity for collective learning that builds from one generation to the next (Christian 2003). The counter image of humans collectively as the potential cause of global-scale social and ecological catastrophe or collapse is a vision of human resourcefulness, restraint and ingenuity as the key to avoiding those dire consequences (Steffen et al. 2007). Although it is now associated with a crisis in human-Earth relations, the Anthropocene must remain flexible enough to evolve into being associated with a transformation to an enduring, mutually enhancing human-Earth relationship (Berry 1999). Indeed, a strong focus on the moral choice that underlies such a possible transformation will help Anthropocene research play "the vital role of helping people make the choices that will realize a morally acceptable future" (Ellis and Trachtenberg 2013: 125).

A transdisciplinary, systems-based approach, as taken in this book, aligns with such a conception of the Anthropocene. At the same time, within this transdisciplinary approach, "[t]he legal system and the strength and weakness of regulatory policies are especially critical factors in the evolution of the Anthropocene" (Ellis and Trachtenberg 2013: 124). Together with the planetary boundaries framework with which it shares signature traits, the Anthropocene draws attention to the aggregate global accumulation of temporally and spatially dispersed impacts caused by temporally and spatially dispersed drivers, an area where the law has yet to find comprehensive and effective solutions (Garver 2013; Castree 2014). With these understandings, the Anthropocene, which is presumed here to have begun within the last 250 years, is a useful frame of reference for highlighting how humanity's unique intellectual capacities both gave rise to the current human dilemma and also must certainly be the foundation of any successful responses to it.

Note

1 The notion of a "good" Anthropocene sometimes aligns with an eco-modernist vision of a complete yet benign taming of nature by humankind (Horn and Bergthaller 2020: 30). This vision rests on human-nature dualism wholly at odds with ecological law.

References

Anthropocene Working Group. 2019. Results of binding vote by AWG, release 21st May 2019. Subcommission on Quaternary Stratigraphy [online]: http://quaternary.stratigraphy.org/working-groups/anthropocene/

Berkhout, F. 2014. Anthropocene futures. *The Anthropocene Review* 1(2): 154–159.

Berry, T. 1999. *The Great Work: Our Way Into the Future*. New York: Three Rivers Press.

Biermann, F. 2014. The Anthropocene: a governance perspective. *The Anthropocene Review* 1(1): 57–61.

Burke, E., III, and K. Pomeranz, eds. 2009. *The Environment and World History*. Los Angeles: University of California Press.

Carpenter, S.R., and E.M. Bennett. 2011. Reconsideration of the planetary boundary for phosphorus. *Environmental Research Letters* 6: 1–12.

Castree, N. 2014. The Anthropocene and geography I: the back story. *Geography Compass* 8(7): 436–449.

Christian, D. 2003. World history in context. *Journal of World History* 14(4): 437–458.

Corlett, R.T. 2015. The Anthropocene concept in ecology and conservation. *Trends in Ecology & Evolution* 30(1): 36–41.

Crutzen, P.J. 2002. Geology of mankind. *Nature* 415: 23.

Crutzen, P.J., and E.F. Stoermer. 2000. The Anthropocene. *Global Change Newsletter* 41: 17–18.

Davies, G. 2002. *A History of Money from Ancient Times to the Present Day*. Cardiff: University of Wales Press.

Economics for the Anthropocene. 2016. *Economics for the Anthropocene* [online]: http://e4a-net.org.

Ellis, E.C. 2011. Anthropogenic transformation of the terrestrial biosphere. *Philosophical Transactions of the Royal Society A* 369: 1010–1035.

Ellis, E.C., and Z. Trachtenberg. 2013. Which Anthropocene is it to be? Beyond geology to a moral and public discourse. *Earth's Future* 2: 122–125.

Fischer-Kowalski, M., F. Krausmann and I. Pallua. 2014. A sociometabolic reading of the Anthropocene: modes of subsistence, population size and human impact on Earth. *The Anthropocene Review* 1(1): 8–33.

Foley, S.F., D. Gronenborn, M.O. Andreae, J.W. Kadereit, J. Esper, D. Scholz, U. Pöschl, D.E. Jacob, B.R. Schöne, R. Schreg, A. Vött, D. Jordan, J. Lelieveld, C.G. Weller, K.W. Alt, S. Gaudzinski-Windheuser, K-C. Bruhn, H. Tost, F. Sirocko and P. Crutzen. 2013. The Palaeoanthropocene—the beginnings of anthropogenic environmental change. *Anthropocene* 3: 83–88.

Garver, G. 2013. The rule of ecological law: the legal complement to degrowth economics. *Sustainability* 5(1): 316–337 [online]: www.mdpi.com/2071-1050/5/1/316/htm

Gibbard, P.L., and M.J.C. Walker. 2014. The term "Anthropocene" in the context of formal geological classification. *Geological Society, London, Special Publications* 395: 29–37.

Gibson-Graham, J.K., and G. Roelvink. 2009. An economic ethics for the Anthropocene. *Antipode* 41: 320–346.

Guillaume, B. 2014. Vernadsky's philosophical legacy: a perspective from the Anthropocene. *The Anthropocene Review* 1(2): 137–146.

Gunderson, L.H., and C.S. Holling, eds. 2002. *Panarchy: Understanding Transformations in Human and Natural Systems.* Washington, DC: Island Press.

Holling, C.S., L.H. Gunderson and D. Ludwig. 2002. In Quest of a Theory of Adaptive Change, pp. 3–22 in L.H. Gunderson and C.S. Holling, eds., *Panarchy: Understanding Transformations in Human and Natural Systems.* Washington, DC: Island Press.

Horn, E., and H. Bergthaller. 2020. *The Anthropocene: Key Issues for the Humanities.* New York: Routledge.

Hourdequin, M. 2013. Restoration and history in a changing world: a case study in ethics for the Anthropocene. *Ethics & the Environment* 18(2): 115–134.

Kellie-Smith, O., and P.M. Cox. 2011. Emergent dynamics of the climate-economy system in the Anthropocene. *Philosophical Transactions of the Royal Society A* 369: 868–886.

Kim, R.E., and K. Bosselmann. 2013. International environmental law in the Anthropocene: towards a purposive system of multilateral environmental agreements. *Transnational Environmental Law* 2: 285–309.

Kosoy, N., P.G. Brown, K. Bosselmann, A. Duraiappah, B. Mackey, J. Martinez-Alier, D. Rogers and R. Thomson. 2012. Pillars for a flourishing earth: planetary boundaries, economic growth delusion and green economy. *Current Opinion in Environmental Sustainability* 4: 74–79.

Kotchen, M.J., and O.R. Young. 2007. Meeting the challenges of the anthropocene: towards a science of coupled human-biophysical systems. *Global Environmental Change* 17: 149–151.

Kotzé, L.J. 2014. Rethinking global environmental law and governance in the Anthropocene. *Journal of Energy & Natural Resources Law* 32(2): 121–156.

Lewis, S.L., and M.A. Maslin. 2015. Defining the Anthropocene. *Nature* 519: 171–180.

Lövbrand, E., S. Beck, J. Chilvers, T. Forsyth, J. Hedren, M. Hulme, R. Lidskog and E. Vasileiadou. 2015. Who speaks for the future of Earth? How critical social science can extend the conversation on the Anthropocene. *Global Environmental Change* 32: 211–218.

Lövbrand, E., J. Stripple and B. Wiman. 2009. Earth System governmentality: reflections on science in the Anthropocene. *Global Environmental Change* 19: 7–13.

Malm, A., and A. Hornborg. 2014. The geology of mankind? A critique of the Anthropocene narrative. *The Anthropocene Review* 1(1): 62–69.

McLellan, R., L. Iyengar, B. Jeffries and N. Oerlemans, eds. 2014. *Living Planet Report 2014: Species and Spaces, People and Places.* Gland, Switzerland: WWF.

McNeill, J.R. 2000. *Something New Under the Sun: An Environmental History of the Twentieth-Century World.* New York: W.W. Norton & Co.

Minteer, B.A. 2012. Geoengineering and ecological ethics in the Anthropocene. *Bioscience* 62(10): 857–858.

Moore, J.W. 2017. The Capitalocene, part I: on the nature and origins of our ecological crisis. *The Journal of Peasant Studies* 44(3): 594–630.

NRC. 2015a. *Climate Intervention: Carbon Dioxide Removal and Reliable Sequestration.* Washington: National Research Council.

NRC. 2015b. *Climate Intervention: Reflecting Sunlight to Cool Earth.* Washington: National Research Council.

Oldfield, F., A.D. Barnosky, J. Dearing, M. Fischer-Kowalski, J. McNeill, W. Steffen and J. Zalasiewicz. 2014. The Anthropocene Review: its significance, implications and the rationale for a new transdisciplinary journal. *The Anthropocene Review* 1(1): 3–7.

Palsson, G., B. Szerszynski, S. Sörlin, J. Marks, B. Avril, C. Crumley, H. Hackmann, P. Holm, J. Ingram, A. Kirman, M. Pardo Buendia and R. Weehuizen. 2013. Reconceptualizing the "Anthropos" in the Anthropocene: integrating the social sciences and humanities in global environmental change research. *Environmental Science & Policy I* 28: 3–13.

Ponting, C. 2007. *A New Green History of the World: The Environment and the Collapse of Great Civilizations.* Toronto: Penguin Books.

Preston, C.J. 2014. Framing an ethics of climate management for the anthropocene, *Climatic Change* [online]: doi:10.1007/s10584-014-1182-4.

Robinson, N. 2014. Fundamental principles of law for the Anthropocene? *Environmental Law and Policy* 44(1–2): 13–27.

Rockström, J., W. Steffen, K. Noone, Å. Persson, F.S. Chapin, III, E. Lambin, T.M. Lenton, M. Scheffer, C. Folke, H. Schellnhuber, B. Nykvist, C.A. De Wit, T. Hughes, S. van der Leeuw, H. Rodhe, S. Sörlin, P.K. Snyder, R. Costanza, U. Svedin, M. Falkenmark, L. Karlberg, R.W. Corell, V.J. Fabry, J. Hansen, B. Walker, D. Liverman, K. Richardson, P. Crutzen and J. Foley. 2009. Planetary boundaries: exploring the safe operating space for humanity. *Ecology and Society* 14(2): 32. [online] URL: www.ecologyandsociety.org/vol.14/iss2/art32/

Ruddiman, W.F. 2013. The Anthropocene. *Annual Review of Earth and Planetary Sciences* 41: 45–68.

Ruhl, J.B. 2012. Panarchy and the law. *Ecology and Society* 17(3): art 31 [online]: http://dx.doi.org/10.5751/ES-05109-170331.

Scott, K.N. 2013. International law in the Anthropocene: responding to the geoengineering challenge. *Michigan Journal of International Law* 34: 309–358.

Slaughter, R.A. 2012. Welcome to the Anthropocene. *Futures* 44: 119–126.

Smith, B.D., and M.S. Zeder. 2013. The onset of the Anthropocene. *Anthropocene* 4: 8–13.

Speth, J.G. 2008. *The Bridge at the Edge of the World.* New Haven, CT: Yale University Press.

Steffen, W., P.J. Crutzen and J.R. McNeill. 2007. The Anthropocene: are humans now overwhelming the great forces of nature. *Ambio* 36(8): 614–621.

Steffen, W., J. Grinevald, P. Crutzen and J. McNeill. 2011. The Anthropocene: conceptual and historical perspectives. *Philosophical Transactions of the Royal Society A* 369: 842–867.

Steffen, W., K. Richardson, J. Rockström, S.E. Cornell, I. Fetzer, E.M. Bennett, R. Biggs, S.R. Carpenter, W. de Vries, C.A. de Wit, C. Folke, D. Gerten, J. Heinke, G.M. Mace,

L.M. Persson, V. Ramanathan, B. Reyers and S. Sörlin. 2015. Planetary boundaries: guiding human development on a changing planet. *Science* 347(6223): 1259885-1–1259885-10.

Testart, A. 1982. The significance of food storage among hunter-gatherers: residence patterns, population densities, and social inequalities. *Current Anthropology* 23(5): 523–537.

Vidas, D. 2011. The Anthropocene and the international law of the sea. *Philosophical Transactions of the Royal Society A* 369.

Visconti, G. 2014. Anthropocene: another academic intervention? *The Two Cultures* 25(3): 381–392.

Waters, C.N., J. Zalasiewicz, C. Summerhayes, A.D. Barnosky, C. Poirier, A. Gałuszka, A. Cearreta, M. Edgeworth, E.C. Ellis, M. Ellis, C. Jeandel, R. Leinfelder, J.R. McNeill, D. deB. Richter, W. Steffen, J. Syvitski, D. Vidas, M. Wagreich, M. Williams, A. Zhisheng, J. Grinevald, E. Odada, N. Oreskes and A.P. Wolfe. 2016. The Anthropocene is functionally and stratigraphically distinct from the Holocene. *Science* 351(6269): aad2622-1–aad2622-10.

Waters, C.N., J.A. Zalasiewicz, M. Williams, M.A. Ellis and A.M. Snelling. 2014. A stratigraphical basis for the Anthropocene? *Geological Society, London, Special Publications* 395: 1–21.

Westley, F., P. Olsson, C. Folke, T. Homer-Dixon, H. Vredenburg, D. Loorbach, J. Thompson, M. Nilsson, E. Lambin, J. Sendzimir, B. Banerjee, V. Galaz and S. van der Leeuw. 2011. Tipping toward sustainability: emerging pathways of transformation. *AMBIO* 40: 762–780.

Zalasiewicz, J., C.N. Waters, C.P. Summerhayes, A.P. Wolfe, A.D. Barnosky, A. Cearreta, P. Crutzen, E. Ellis, I.J. Fairchild, A. Gałuszka, P. Haff, I. Hajdas, M.J. Head, J.A. Ivar do Sul, C. Jeandel, R. Leinfelder, J.R. McNeill, C. Neal, E. Odada, N. Oreskes, W. Steffen, J. Syvitski, D. Vidas, M. Wagreich and M. Williams. 2017. The working group on the Anthropocene: summary of evidence and interim recommendations. *Anthropocene* 19: 55–60.

Zalasiewicz, J., M. Williams, R. Fortey, A. Smith, T.L. Barry, A.L. Coe, P.R. Bown, P.F. Rawson, A. Gale, P. Gibbard, F.J. Gregory, M.W. Hounslow, A.C. Kerr, P. Pearson, R. Knox, J. Powell, C. Waters, J. Marshall, M. Oates and P. Stone. 2011a. Stratigraphy of the Anthropocene. *Philosophical Transactions of the Royal Society A* 369: 1036–1055.

Zalasiewicz, J., M. Williams, A. Haywood and M. Ellis. 2011b. The Anthropocene: a new epoch of geological time? *Philosophical Transactions of the Royal Society A* 369: 835–841.

2 The human dilemma in the Anthropocene

Earth's life systems currently face a predicament without precedent, due to an expanding aggregation of anthropogenic impacts that by many measures has already overshot global ecological limits (Ponting 2007; Rockström et al. 2009; Carpenter and Bennett 2011; Steffen et al. 2011; Berkhout 2014; McLellan et al. 2014; Steffen et al. 2015). Concern over the extent of human manipulation of material and energy stocks and flows and the resulting consequences for the global ecosystem, and the trajectory of current trends in this pattern of social metabolism, gave rise to the term Anthropocene.

The current predicament for Earth's life systems has subsidiary features that compound its severity, particularly the intergenerational and intragenerational distribution of the means for well-being among humans and between humans and other species. A significant and growing rift separates the world's richest and poorest people (Piketty 2014). In 2014, the world's wealthiest 1 percent of people owned an estimated 48 percent of total wealth and the bottom 80 percent owned only 5.5 percent (Hardoon 2015). In 2010, 925 million people, mostly in Sub-Saharan Africa and Asia, were chronically hungry (UN-FAO 2010) and up to 4.5 billion people suffered some form of nutritional deficiency (Schade and Pimentel 2010). At the same time, more than a billion people are now "overfed, overweight and obese" (Schade and Pimentel 2010: 246). Moreover, the proportion of ecological harm that poor countries suffer is higher than their proportionate contribution to the drivers of those impacts (Ponting 2007; Srinivasan et al. 2008).

As for other species, human activities have had wide-ranging impacts on their abundance, diversity and global distribution for millennia, and especially since the dawn of the era of fully global trade around 1500 (Ponting 2007). The populations of some species, like domesticated plants and animals, have grown vastly in the course of history (McNeill 2000), while the abundance of most has declined; at the beginning of the twenty-first century, humans make up about five percent of the total animal mass of the planet (McNeill 2000). According to one study, the total population of non-human vertebrate species decreased by about half between 1970 and 2010 (McLellan et al. 2014). Humans may have enhanced the rate of species extinctions by a factor of between 100 and 1000 (Rockström et al. 2009), and the current dilemma is on track to lead to the Earth's sixth mass extinction (Ponting 2007; Kolbert 2014).

Ultimately, the current dilemma for humanity is in being caught between two irreconcilable narratives of impossibility regarding the human prospect in the Anthropocene, both of which derive from the trajectory of the human-Earth relationship from the first humans to the current era (Garver 2013). One narrative says it is impossible to improve human and social welfare, or to achieve "sustainable development" under prevailing definitions,[1] unless the human economy (measured by the inflation-adjusted monetary value of goods and services exchanged in the economy) grows indefinitely. This is the growth-insistent narrative. The other says it is impossible to avoid impacts on the global ecosystem that ultimately would be catastrophic to present and future generations of human and other life if the global economy grows indefinitely, and that a downsizing and stabilization of the economy is necessary to reverse global ecological trends and restore degraded ecological functions. This is the limits-insistent narrative.

The growth-insistent narrative remains politically unassailable around the world. Most people and the law and policymakers and other decision-makers they support, elect or are controlled by assume that conditions they find desirable and associate with happiness and well-being—such as increased levels of personal ease and comfort, advances in medicine, increased mobility, a limited range of pollution control, an extraordinary and growing array of consumer choices—are only possible with a growing economy. Historical correlation provides some support for this view, but only to a point (Stiglitz et al. 2009) and without proving growth causes these attributes. According to the growth-insistent narrative, economic growth (typically measured by gross domestic product, GDP) is necessary to create enough work for rising populations as labor productivity increases, as well as to solve problems of poverty and pollution. In this view, aggregate ecological impacts of the global economy as it serves a growing human population can be handled adequately by perpetually increasing the efficiency of the throughput of material and energy in the economy (Giljum et al. 2008; UNEP 2011).

Ethically, under many versions of the growth-insistent narrative, self-interested behavior, profit making, wealth maximization and the distributive workings of markets are assumed overall to ensure fairness and justice implicitly (Sen 1999). In another version, they are considered a necessary evil until all people are lifted out of poverty and into the good life (Skidelsky and Skidelsky 2012)—a vague horizon that tends to loom always in the distant future. Ethics of sharing (interhuman, interspecies and intergenerational), sufficiency and "right relationship" within the commonwealth of life are sometimes acknowledged and given moral weight, but they have a far too limited role in the key normative regimes, like law, that shape the overall direction of human societies (Brown 2008; Brown and Garver 2009; Skidelsky and Skidelsky 2012). A mutually enhancing human-Earth relationship provides a more sound foundation for these ethics.

Scientific evidence of the mounting aggregate ecological disruption that correlates with economic growth is seriously undermining the growth-insistent narrative. The conceptual foundation of "safe operating space for humanity" within planetary boundaries (Rockström et al. 2009; Steffen et al. 2015) aligns with the limits-insistent narrative (Garver 2013; Kim and Bosselmann 2013) because global ecological boundaries, not insistence on economic growth, constrain the socio-political and economic spheres. "The thresholds in key Earth System processes exist irrespective of peoples' preferences, values or compromises based on political and socioeconomic feasibility" (Rockström et al. 2009: 7). The ever-increasing amount of material and energy that accumulates in or passes through the exponentially growing human economy, or is ecologically degraded or destroyed by human domination of land and water, places a mounting strain on ecosystems that serve as both sources and sinks of human appropriation.

Clear evidence shows a close correlation between the rise in metrics of human pressure on Earth's ecosystems, such as human population, economic growth and human use of material and energy, and global ecological change, such as global heating, loss of biodiversity and excessive nutrient loading of ecosystems (McNeill 2000: Table 12.1, 360–361; Speth 2008; Steffen et al. 2011). If the global economy grows at two percent to three percent, rates generally assumed or hoped for in key international contexts (Stern 2007; IPCC 2007), it will be roughly 9 to 16 times bigger in 2100 than in 2000. Thus, by 2100, absurd levels of decoupling of the throughput of material and energy in the economy (that is, the use of material and energy per unit of Gross World Product) from the economy's ecological impacts would be necessary for the human enterprise to stay within global ecological limits it has already begun to transgress (Kallis 2018). The prognosis of economists who defend their insistence on perpetual growth with confident optimism about the ability of yet-unknown technologies to yield the necessary perpetual exponential increase in resource productivity is factually flawed (CBC Radio 2011; UNEP 2011). The limits-insistent narrative has not yet amassed sufficient popular support at any significant level of government, but it has given rise to growing academic, professional and social studies (Victor 2008; Jackson 2009; Rockström et al. 2009; Steffen et al. 2015; Kothari et al. 2015), movements and associations, such as various ecological economics professional societies and the degrowth movement (Research & Degrowth 2016; Kallis 2018).

The march toward this clash of narratives that defines the key challenge of the Anthropocene corresponds to the long history of expansion and growth in the presence of humans on Earth and of impacts of humanity on the global panarchy. Although exceptions may be found, expansion and "developmentalism" (Ponting 2007; Pomeranz 2009) that transcend time, culture and socio-political paradigms are the hallmarks of human history leading to the Anthropocene. The globally dominant growth-insistent narrative emerged from this historical evolution.

Although the current state of human-Earth relations urgently calls for a shift to global socio-political arrangements founded on the limits-insistent narrative, the growth-insistent narrative has a powerful and unrelenting grip on the dominant machinery of global human affairs. From the global to the local level, legal systems inevitably are fully integrated into the panarchy of complex dynamic systems that underlie the emergence of the Anthropocene. These legal systems remain lodged in a mechanistic worldview that came to dominance in the Enlightenment—a worldview that assumes non-human nature is subject to human dominance and control through ownership, relies on linear cause-and-effect relationships and adopts the reductionist view that measuring and quantifying the discrete elements of a system yields a full understanding of the whole (Capra and Mattei 2015). Ultimately, how human society crafts its legal systems from the local to the global level will influence the potential for anthropogenic impacts to trigger globally or regionally catastrophic shifts in the economic, political, social and ecological systems with which law interacts and evolves. The following sections describe the historical trajectory and key features of the dilemma that the Anthropocene currently poses to humanity and examines the role of law and legal systems in the emergence of the Anthropocene.

Note

1 Principle 12 of the 1992 Rio Principles, the United Nations Environment Programme's green growth strategy, the outcome document of the Rio+20 conference and the United Nations' Sustainable Development Goals all treat economic growth as a non-negotiable element of sustainable development (UNEP 2011; United Nations 2012; United Nations 2015).

References

Berkhout, F. 2014. Anthropocene futures. *The Anthropocene Review* 1(2): 154–159.

Brown, P.G. 2008. *The Commonwealth of Life: Economics for a Flourishing Life, 2nd Edition.* Montreal: Black Rose Books.

Brown, P.G., and G. Garver. 2009. *Right Relationship: Building a Whole Earth Economy.* San Francisco: Berrett-Koehler.

Capra, F., and U. Mattei. 2015. *The Ecology of Law: Toward a Legal System in Tune with Nature and Community.* San Francisco: Berrett-Koehler.

Carpenter, S.R., and E.M. Bennett. 2011. Reconsideration of the planetary boundary for phosphorus. *Environmental Research Letters* 6: 1–12.

CBC Radio. 2011. *Green Growth or No Growth* [online]: www.cbc.ca/ideas/episodes/2011/02/02/green-growth-or-no-growth/

Garver, G. 2013. The rule of ecological law: the legal complement to degrowth economics. *Sustainability* 5(1): 316–337 [online]: www.mdpi.com/2071-1050/5/1/316/htm

Giljum, S., A. Behrens, F. Hinterberger, C. Lutz and B. Meyer. 2008. Modelling scenarios towards a sustainable use of natural resources in Europe. *Environmental Science & Policy* 11: 204–216.

Hardoon, D. 2015. *Wealth: Having It All and Wanting More*. Oxfam Issue Briefing, January 2015. Oxford UK: Oxfam International.

IPCC. 2007. *Climate Change 2007: The Physical Science Basis. Contribution of Working Group I to the Fourth Assessment Report of the Intergovernmental Panel on Climate Change* [S. Solomon, D. Qin, M. Manning, Z. Chen, M. Marquis, K.B. Averyt, M. Tignor and H.L. Miller (eds.)]. New York: Cambridge University Press.

Jackson, T. 2009. *Prosperity Without Growth: Economics for a Finite Planet.* Washington, DC: Earthscan.

Kallis, G. 2018. *Degrowth.* Newcastle upon Tyne, UK: Agenda Publishing.

Kim, R.E., and K. Bosselmann. 2013. International environmental law in the Anthropocene: towards a purposive system of multilateral environmental agreements. *Transnational Environmental Law* 2: 285–309.

Kolbert, E. 2014. *The Sixth Extinction: An Unnatural History.* New York: Henry Holt and Company.

Kothari, A., F. Demaria and A. Costa. 2015. Buen Vivir, Degrowth and ecological swaraj: alternatives to sustainable development and the Green Economy. *Development* 57(304): 362–375.

McLellan, R., L. Iyengar, B. Jeffries and N. Oerlemans, eds. 2014. *Living Planet Report 2014: Species and Spaces, People and Places.* Gland, Switzerland: WWF.

McNeill, J.R. 2000. *Something New Under the Sun: An Environmental History of the Twentieth-Century World.* New York: W.W. Norton & Co.

Piketty, T. 2014. *Capital in the Twenty-First Century.* Cambridge MA: The Belknap Press of Harvard University Press.

Pomeranz, K. 2009. Introduction: World History and Environmental History, pp. 3–32 in E. Burke III and K. Pomeranz, eds., *The Environment and World History.* Los Angeles: University of California Press.

Ponting, C. 2007. *A New Green History of the World: The Environment and the Collapse of Great Civilizations.* Toronto: Penguin Books.

Research & Degrowth. 2016. *Research and actions to consume less and share more* [online]: www.degrowth.org.

Rockström, J., W. Steffen, K. Noone, Å. Persson, F.S. Chapin III, E. Lambin, T.M. Lenton, M. Scheffer, C. Folke, H. Schellnhuber, B. Nykvist, C.A. De Wit, T. Hughes, S. van der Leeuw, H. Rodhe, S. Sörlin, P.K. Snyder, R. Costanza, U. Svedin, M. Falkenmark, L. Karlberg, R.W. Corell, V.J. Fabry, J. Hansen, B. Walker, D. Liverman, K. Richardson, P. Crutzen and J. Foley. 2009. Planetary boundaries: exploring the safe operating space for humanity. *Ecology and Society* 14(2): 32 [online] URL: www.ecologyandsociety.org/vol.14/iss2/art32/

Schade, C., and D. Pimentel. 2010. Population crash: prospects for famine in the twenty-first century. *Environment, Development & Sustainability* 12: 245–262.

Sen, A. 1999. *Development as Freedom.* New York: Anchor Books.

Skidelsky, R., and E. Skidelsky. 2012. *How Much Is Enough? Money and the Good Life.* New York: Other Press.

Speth, J.G. 2008. *The Bridge at the Edge of the World.* New Haven, CT: Yale University Press.

Srinivasan, U.T., S.P. Carey, E. Hallstein, P.A.T. Higgins, A.C. Kerr, L.E. Koteen, A.B. Smith, R. Watson, J. Harte and R.B. Norgaard. 2008. The debt of nations and the distribution of ecological impacts from human activities. *Proceedings of the National Academy of Science* 105(5): 1768–1773.

Steffen, W., J. Grinevald, P. Crutzen and J. McNeill. 2011. The Anthropocene: conceptual and historical perspectives. *Philosophical Transactions of the Royal Society A* 369: 842–867.

Steffen, W., K. Richardson, J. Rockström, S.E. Cornell, I. Fetzer, E.M. Bennett, R. Biggs, S.R. Carpenter, W. de Vries, C.A. de Wit, C. Folke, D. Gerten, J. Heinke, G.M. Mace, L.M. Persson, V. Ramanathan, B. Reyers and S. Sörlin. 2015. Planetary boundaries: guiding human development on a changing planet. *Science* 347(6223): 1259885-1–1259885-10.

Stern, N. 2007. *The Economics of Climate Change: The Stern Review.* Cambridge UK: Cambridge University Press.

Stiglitz, J., A. Sen and J.P. Fitoussi. 2009. *Report by the Commission on the Measurement of Economic Performance and Social Progress* [online]: www.stiglitz-sen-fitoussi.fr/documents/rapport_anglais.pdf

United Nations. 2012. *The future we want.* New York: United Nations [online]: https://documents-dds-ny.un.org/doc/UNDOC/GEN/N11/476/10/PDF/N1147610.pdf?OpenElement.

United Nations. 2015. *Transforming Our World: The 2030 Agenda for Sustainable Development.* New York: United Nations.

United Nations Environment Programme (UNEP). 2011. *Towards a Green Economy: Pathways to Sustainable Development and Poverty Eradication—A Synthesis for Policymakers*. Nairobi: UNEP. www.unep.org/greeneconomy/GreenEconomyReport/tabid/29846/language/en-US/Default.aspx

United Nations Food and Agriculture Organization (UN-FAO). 2010. Global hunger declining, but still unacceptably high. Rome: UN-FAO.

Victor, P. 2008. *Managing Without Growth.* Cheltenham, UK: Edward Elgar.

3 Energy transitions in history and their impacts

"Humankind has long been part of nature, but a distinctive part, especially in recent millennia, when it acquired the power and the numbers to become a rogue mammal, exerting ever-increasing influence upon earthly ecosystems" (McNeill 2003: 6). The human traits and tendencies that have given way to the Anthropocene began as early as the Pleistocene, continued with the Neolithic Transition, proliferated at an accelerated pace once humans took to harnessing the energy of fossil fuels and has surged even more since the mid-twentieth century (Ponting 2007). The Anthropocene is the result of a continuum of human activities and energetics that preceded the extensive use of fossil fuels starting around 1800, such as hunting and gathering, controlled use of fire, development of increasingly sophisticated tools and technologies, development of symbolic language, storage of commodities, agriculture, use of money, specialization, social hierarchies, long-distance commerce, colonization and urbanization (Smil 2007; Ponting 2007). Whatever recent geological markers of an overarching impact of humans on the Earth might emerge to define the Anthropocene formally, any suggestion of too sharp a transition along this continuum would be deceptive. The fossil fuel era has merely accelerated the extent and expanded the nature of human impacts on Earth's systems.

The networks of relationship among humans in different parts of the world, at all scales, and among humans and Earth systems are increasingly complex. The interconnecting of human societies through complex globalized systems founded on increasing industrialization, urbanization and trade is a relatively new feature of human-Earth systems (Ponting 2007). The historical legacy on which the current predicament rests is loaded with a sweeping complex of anthropogenic material and energy flows, population growth and dispersal, physical and institutional infrastructure, patterns of behavior, investment-backed expectations and ideological perspectives that as an integrated whole has enormous momentum and inertia, which in many ways will resist change (Oldfield et al. 2014). No single factor was sufficient alone to give rise to the current global ecological dilemma.

Studies within the rubric of environmental history are a rich source for examining the emergence of the Anthropocene. Because of its reliance on other disciplines, in particular ecology and earth systems science, environmental history tends to take a systems view in which the human-Earth relationship is infused with multiple feedbacks and non-linear co-evolution of multiple variables (Cronon 1993; McNeill 2000; McNeill 2010a). Environment history "bind[s] together approaches from the social sciences, the natural sciences and history in non-deterministic accounts of change" (Sörlin and Warde 2007: 118). Factors giving rise to the Anthropocene and the current human dilemma therefore should be seen as co-evolutionary with each other (McNeill 2010b). They encompass a set of interrelated forces of global importance, even though their manifestation has varied significantly both temporally and spatially.

Material and energetic factors of the emergence of the Anthropocene relate to human use of materials derived from the living and non-living environment. Four interrelated material factors that have been particularly important in bringing humanity to the current predicament are: (1) human use and development of tools and technology, (2) agriculture, (3) human use of fossil fuels, and (4) growth and spread of the human population. The transitions from hunter-gatherer societies to agricultural societies and from agricultural societies to industrial ones are key milestones in the environmental history of these four factors.

The legacy of early humans: technology, control of fire and language

Before the emergence of agriculture that occurred independently in different parts of the world beginning around 10,000 years ago, human subsistence was based almost entirely on hunting and gathering. The direct ancestors of modern humans began making stone tools at least two million years ago, and the full tool kit for early humans eventually included wooden spears, fire, animal skins and tools made from bone and ivory (Ponting 2007). In this long period of human development, which lasted roughly until 10,000 years ago, the emergence of complex language and increasingly sophisticated tools distinguished humans more and more from other species (Ponting 2007). Although forays out of Africa began at least 700,000 years ago, permanent human presence in Asia, Europe and beyond gradually took hold between 80,000 and 20,000 years ago (Ponting 2007). Humans reached the Americas no later than about 13,000 years ago and rapidly spread, such that humans were present in all major parts of the world by about 10,000 years ago (Ponting 2007).

Tools, fire and language likely facilitated the ability of humans to adapt to new ecological contexts, from tropical to glacial, as they spread across Earth. Although human populations were generally sparsely distributed and nomadic prior to the Neolithic Transition, humans altered the ecosystems they inhabited with fire to create clearings or to assist in hunting, preferential treatment of favored plants and hunting that affected wildlife populations (Foley et al. 2013). In some regions, hunting prior to the Neolithic Transition led to extinctions of some species and significant decreases in others, especially large mammals (Ponting 2007; Foley et al. 2013).

The astonishing progression from the earliest controlled use of fire to the modern use of fossil fuels, from early forms of hunting and gathering with tools to modern industrial modes of provisioning and from the first symbolic forms of language to modern systems of information management and communication is readily apparent. Looking into the future, plausible trajectories include implementation of global-scale geoengineering projects, nanotechnologies and artificial intelligence technologies approaching or surpassing human intelligence (the so-called "singularity") (Robock 2008; Harari 2014). A detailed description of the vast evolution up to the contemporary era of early humans' development of tools, fire and language, which the human capacity for transgenerational collective learning made possible, is beyond the scope of this book. However, further descriptions of key aspects of this evolution and its impacts on the human-Earth relationship are included in the following sections.

Agriculture

The Neolithic Transition that began around 10,000 years ago may well be "the most profound leap in the history of humanity" (Sieferle 2001: 9). Fischer-Kowalski and Haberl (1997) identify "colonization" as the key to the transition from hunting and gathering to agrarian lifestyles. Colonization is "the complex of social activities that deliberately change important parameters of natural systems and actively maintain them in a state different from the conditions that would prevail in the absence of human interventions" (Fischer-Kowalski and Haberl 1997: 64–65). The main form of colonization that gave rise to the Neolithic Transition is agriculture, which arose independently in Africa, Asia, Oceania and the Americas (Smil 2007; Ponting 2007; Dow et al. 2009; Foley et al. 2013). Whereas hunter-gatherers (even with basic tools and fire) generally tapped into ecosystemic energy flows without systematically reconfiguring or controlling them, Neolithic farmers initiated the systematic alteration of ecosystems (that is, colonization) to increase their energetic value for human uses (Fischer-Kowalski and Haberl 1997; Sieferle 2001).

A cascade of ecosystemic and socio-economic effects unfolded from the onset of agriculture. Farming promoted a significant rise in fixed human settlements and social complexity, which existed in hunter-gatherer societies but generally only in highly productive environments, especially where long-term food storage was possible (Testart 1982; Pryor 1986; Smil 2007). The diet of the original farmers generally was nutritionally poorer and less varied than that of the hunters and gatherers amid whom they settled (Sieferle 2001; Smil 2007; Ponting 2007), and agrarians faced new and more widespread risks of famines if crops failed. Nonetheless, a key advantage of the sedentary agrarian lifestyle was enhanced storage of food—a way to "bundle the energy flow" to support more people, with less uncertainty (Sieferle 2001: 13–14; Smil 2007).

In ancient Egypt, with similar patterns elsewhere, large-scale storage of grain is associated with the earliest signs of social stratification. Those who controlled grain stores and their distribution achieved higher wealth and social status (Testart 1982; Davies 2002; Ponting 2007). Agriculture allowed for the emergence of an elite, typically from five to ten percent of the population prior to the

Industrial Revolution, who were not food producers but instead relied on food surpluses (Ponting 2007). Social stratification and role specialization gave rise to increasingly complicated barter systems, and eventually, when the number of tradeable goods reached some threshold where bartering became impractical (Davies 2002), to the first forms of commodity money. In the shift from gift economies of hunter and gatherer societies to early banking and monetary systems in ancient agriculture-based civilizations, "property and power relationships took the place of kinship relationships as the central form of organising social cohesion" (Sieferle 2001: 12). "Agriculture stimulated the spread of specialized crafts, inequality, hierarchy, and organized warfare, and was necessary for the emergence of cities, states, writing, and ultimately modern civilization itself" (Dow et al. 2009).

Agriculture is the root of many forms of ecologically disruptive human behavior that strains scarce resources (renewable and non-renewable) or overwhelms the capacity of ecosystems to handle the outputs of human transformations of material and energy. Agriculture is a form of simplification and intensification of land use (Scott 1998) that, even in its most primitive forms, alters ecosystems it displaces. Because agriculture almost always reduces species diversity, and harvest reduces the energy available for life support, farming tends inherently to reduce biodiversity (Haberl et al. 2009). Strategies for meeting the basic challenge of finding, creating or maintaining suitable soil and moisture conditions for agriculture include swidden agriculture, fallow periods, crop rotation and multi-cropping, no-till cultivation, irrigation and nutrient addition. All have ecological implications, which may include loss or degradation of habitat for terrestrial and freshwater species, reduced capacity for carbon storage, erosion and salinization of soil, desertification and water and air pollution from nutrient and pesticide run-off and drift (Jacobsen and Adams 1958; McNeill 2000; Richards 2003; Rengasamy 2006; Smil 2007; Sörlin and Warde 2007).

The environmental history of agriculture during the early modern period (roughly, 1400 to 1800) in different parts of the world is replete with accounts of its significant ecological impacts. Reliance on energy from biomass, wind and water placed a strong demand on agriculture, not only for food and fiber for human survival, but also the energy needed to extract work from humans and animals that powered much of the ever-expanding economy (Sieferle 2001; Smil 2007; Ponting 2007). The basic strategies were to intensify agricultural production locally with increased inputs of labor or other types of energy, irrigation or modified production methods; or to augment production extensively by moving to or relying (through trade or conquest) on external areas, either nearby (e.g. drained wetlands) or remote (e.g. the Baltics, Eastern Europe, Scandinavia, the Americas, northern China, Taiwan, South Africa) (Hoffman 2001; Pomeranz 2002; Jones 2003; Richards 2003; Smil 2007). In the late medieval and early modern era, metabolic rift (i.e. the removal of some portion of bioregional metabolism to remote areas) and increasing reliance on "ghost acres" (i.e. the amount of land spared for agricultural or other uses by using remote lands or new energy regimes) became firmly established (Hoffman 2001; Pomeranz

2002; Jones 2003; Clark and Foster 2009). Through expanding trade, people, especially in wealthy industrial societies, have become increasingly detached from the ecosystems that maintain them; based on consumption, individual ecological footprints globally are less and less local (Nyquist 2013). Ongoing patterns of land and resource "grabbing" exacerbate these trends (Pearce 2012). Land and resource grabbing occurs when remote investors drive conversion of land and resources from generally small-scale, traditional, locally-controlled uses or relatively undeveloped wilderness to massive holdings devoted to industrial agriculture or resource extraction serving global markets of finance, goods and services. Typically, these remote owners know little about the people and ecosystems their decisions affect significantly and mostly adversely; they are simply looking for high returns on their investment (Garver 2019).

With the increase in long-distance oceanic voyages starting around 1500, European countries brought new forms of intensive agricultural land use to the Americas and elsewhere. For example, Caribbean sugar plantations in the seventeenth and eighteenth centuries stand out as a new agricultural form with massive local ecosystemic consequences, as well as all the social consequences associated with colonialism and the African slave trade (Richards 2003). Europeans brought Old World domesticated plants and animals to the Americas (e.g. wheat, sheep, horses and cattle) and introduced New World domesticated plants (e.g. maize, tomatoes and potatoes) to the Old World (Crosby 1987; Richards 2003). They also brought Old World diseases, most notably smallpox, which killed off 50 percent or more of Indigenous populations and thereby temporarily reduced the ecosystemic impacts of Indigenous agriculture (Crosby 1987; Richards 2003). In return, the Columbian Exchange brought syphilis from the Americas to Europe (Ponting 2007).

By the end of the early modern era, a world market was well established, creating an unprecedented level of global interconnectedness of agricultural production and its cascade of impacts (Beddoe et al. 2009). The modern globalized economy is the progeny of this market. Yet, agriculture and trade prior to about 1800 were still contained within the energetic limits on transport and production methods achievable in an agrarian energy system (Sieferle 2001; Smil 2007). The nature and ecological impacts of modern agriculture, transformed over the past two centuries by fossil fuels and the scientific, industrial and chemical revolutions, are of another order altogether. Fossil-fueled modern agriculture depends on mechanization, automation, massive use of inorganic fertilizers and pesticides, irrigation that relies on major alterations of aquatic and hydrologic systems with dams and other water management infrastructure, an industrial approach that favors disease-vulnerable monocultures of a dwindling number of species of plants and animals, genetic engineering, increasing consumption of meat, and exacerbation of the metabolic rift between where food and fiber are produced and consumed (McNeill 2000; Johnson 2006; Smil 2007; Clark and Foster 2009; Rockström et al. 2009; Foley et al. 2011).

Modern agriculture contributes significantly to climate change, disruption of global nutrient cycles, biodiversity loss, land use change and other pressures on planetary boundaries. Since 1800, cropland has tripled; it now takes up about 12 percent of the Earth's ice-free land surface, and pastures another 22 percent—a vast transformation since the Neolithic Transition (McNeill 2000; Ramankutty et al. 2008; Rockström et al. 2009). Irrigated land increased from 75 million hectares at the end of World War II to over 275 million hectares in 2000, mostly in Asia (Smil 2007: 294). Agriculture globally places significant pressure on accessible freshwater supplies (Ramankutty et al. 2008; Rockström et al. 2009). Crop farming uses about five percent of total primary energy supply globally, requiring fossil fuel inputs at every stage (Smil 2007). Methane from rice paddies and livestock, nitrous oxide emissions from nitrogen fertilization and nitrification processes and basic loss of carbon storage capacity also contribute to climate change (Ramankutty et al. 2008; Rockström et al. 2009). Anthropogenic conversions of atmospheric nitrogen to reactive forms that enter ecosystems were virtually non-existent before the industrial era, but are now equal to natural conversions, doubling the global load (Rockström et al. 2009; Steffen et al. 2015). Phosphorus, which unlike nitrogen derives from exhaustible fossil minerals, is also heavily used as an agricultural fertilizer. Anthropogenic phosphorus and nitrogen loads contribute to eutrophication of freshwater ecosystems and oxygen-depleted dead zones where surface run-off enters ocean waters (Rockström et al. 2009; Carpenter and Bennett 2011; Steffen et al. 2015).

Human use of fossil fuels

Local exhaustion or diminishment of both renewable and non-renewable resources has a long history that predates the age of fossil fuels and was a general feature of the early modern era that led up to the Industrial Revolution (Sieferle 2001; Richards 2003; Diamond 2005). Although whales, forests and plantation slaves gained some relief when fossil fuels provided alternatives to whales for bone, oil and wood for fuel and humans for labor, the current human predicament would not exist without the dramatic shift to fossil fuels in the past two centuries. Indeed, this is a key reason why the Anthropocene most often refers to the period beginning with the onset of the Industrial Revolution in the late eighteenth century.

Fossil fuels allowed a fundamental shift from the "basic metabolism" in an agrarian system to "extended metabolism," which involves use of non-renewable resources such as fossil fuels and metal and mineral ores to augment total social metabolism (Fischer-Kowalski and Haberl 1997). With the consequent release of ancient stores of material and high-order energy into biospheric systems and their cycles, "the most immediate sustainability problems shift from resource supply to the bottleneck of pollution tolerance" (Fischer-Kowalski and Haberl 1997: 63). The scale and intensity of a human enterprise dependent on fossil fuels has now pushed this tolerance perilously close to, or perhaps past, the breaking point (McNeill 2000; Rockström et al. 2009; Steffen et al. 2015).

The nineteenth century, at least in economically more powerful countries, was a period of rapid transition from a solar energy society to one powered primarily by coal (and later oil and gas), with labor moving in droves from the land to factories. By contrast, the twentieth century was a period during which the ripple effects of using these new energy sources took off explosively (McNeill 2000; Ponting 2007; Pomeranz 2009). The overarching ecological effect of fossil fuel use in the twentieth century was to increase the scale and intensity of pre-industrial environmental impacts. However, many toxic chemicals, substances that deplete stratospheric ozone and risks associated with nuclear energy, biotechnology and nanotechnology were new (McNeill 2000; Smil 2007). All the mutually reinforcing sectors of the economy—agriculture, buildings, energy supply, fisheries, forestry, industry, transportation, waste and water (UNEP 2011)—gained a significant fossil fuel boost. "Modern, high-energy civilization—marked by megacities, globalized economy, unprecedented levels of affluence, intensive transportation, instant communication, a surfeit of food, and the amassment of possessions—could not have arisen without high energy densities of fossil fuels, portability of refined oil products, and superior flexibility of electricity" (Smil 2007: 380). As well, the vast increase in synthetic chemicals and advances in medicine have allowed a doubling or more of global life expectancy at birth in the twentieth century (United Nations 1999; Olshansky et al. 2001; United Nations 2010). Underlying all these features of modern life is the artificialization, mechanization, urbanization and industrialization that fossil fuels also greatly enhance (McNeill 2000).

The explosion of material and energy use in the fossil fuel era correlates strongly with rising ecological impacts from that use, as well as with rising GDP and human population (Krausmann et al. 2009; Steffen et al. 2011). Among other things, the fossil fuel boost has allowed expanded access to other exhaustible resources, like phosphates and other minerals, fossil water and metal ores with declining metal content. Fossil fuels provided relief from the limitations that wood supply or water power placed on the returns from low-grade or hard-to-access metal ores, and allowed the enormous ecological disruption associated with mining and other resource extraction to expand significantly (McNeill 2000; Sieferle 2001; Moore 2007).

The surge of energy that coal, oil and natural gas have supplied has given way to exponential growth in human population and to the throughput of material and energy in human society. Yet, these energy resources are limited, and the energy return on the energy invested (EROEI) to exploit them is steadily decreasing (Smil 2007; Hall and Klitgaard 2012). The same is true for other material resources, like phosphate deposits, fossil water and metals, the extraction of which in turn relies heavily on energy from fossil fuels. Parallel to this one-way, non-cyclical depletion of material and energy sources that have undergirded societal development increasingly in the last two centuries, the by-products of the throughput of material and energy sources in society have had and continue to have profound impacts on Earth systems. Fossil-fueled human society is profoundly changing the structure, function and chemical composition of the atmosphere; stratosphere and terrestrial and aquatic systems; and the abundance (both totally and relatively) and diversity of life. According to research on planetary boundaries, these changes may become catastrophic to human and other life on Earth. Indeed, their momentum and the feedbacks associated with them are already to some extent unstoppable and irreversible. Whether humanity can avoid crucial tipping points in Earth's life support systems, more imminent with exploitation of fossil fuels, is the critical question of this era.

Increase and dispersal of the human population

Homo sapiens, a species representing about 5 percent of the total animal mass of the planet (McNeill 2000), reached 7.7 billion in number in 2019 (United Nations 2019). The IPAT framework (Ehrlich and Holdren 1972) includes population size as a critical factor in considering aggregate human impact, however defined. For example, the human impact (I) on biodiversity depends on the total human population (P), the per capita metabolism of material and energy in society (A, affluence, better thought of as metabolism or consumption) and the human technologies (broadly construed) that affect biodiversity (T, which could be expressed as biodiversity impact per unit of metabolism). Yet, the IPAT framework masks a good deal of complexity, as well as heterogeneous dispersion of both drivers and impacts across spatial and temporal scales (Kallis 2018: 182). Lifestyles and ethical choices matter, and a historical perspective suggests that overconsumption by a small percentage of the human population may be more ecologically problematic than rapid population growth among poorer people who consume much less (Smil 2007; Giljum et al. 2008; Brown and Garver 2009; United Nations 2011).

Human numbers have had temporary down ticks and regional variation, as with major plague-ridden periods in the fourteenth century and earlier in Europe (Ponting 2007; Burke and Pomeranz 2009) and massive loss of human life due to the vulnerability of Indigenous Americans to European diseases such as smallpox (Ponting 2007) in the early modern era. However, the human population has generally grown throughout history, along with associated aggregate ecological impacts. After the Neolithic Transition, the rise was slow and erratic until the jump in long-distance commerce beginning around 1500, with an even more radical acceleration during the Industrial Revolution, especially in the twentieth century (McNeill 2000). Both the Neolithic Transition and Industrial Revolution took place despite important disadvantages. Agriculture emerged despite poorer nutrition, more infectious diseases, increased human labor inputs per unit of food energy and increased vulnerability to famine (Armelagos et al. 1991; Sieferle 2001; Smil 2007; Ponting 2007; Burke and Pomeranz 2009). In the case of coal at the dawn of the fossil fuel era, acute pollution and health effects and the challenges of mining were drawbacks (Merchant 1980; Sieferle 2001; Pomeranz 2002; Richards 2003; Ponting 2007). Nonetheless, both transitions relieved pressures on population growth and then enhanced population growth.

Although the origins of agriculture remain unknown, increases in population density and diminishing returns from hunting and gathering are likely factors, along with climatic perturbations during the post-Ice Age transition to favorable climatic conditions that have persisted during the Holocene (Pryor 1986; Sieferle 2001; Smil 2007; Burke 2009; Dow et al. 2009; Rockström et al. 2009). Agriculture allowed population growth to accelerate (Fischer-Kowalski and Haberl 1997; McNeill 2000), and during the agrarian era, the global human population grew from a very rough estimate of 8 million 10,000 years ago, to 250–300 million 2 thousand years ago, about 450 million in 1500, and just under 1 billion in 1800 (Armelagos et al. 1991; Burke and Pomeranz 1999; McNeill 2000; United Nations 2011).

The use of fossil fuels took off in the wake of Malthus's warning that exponential population growth would outpace arithmetic growth in food production, with war, disease and famine, along with education and foresight, perpetually keeping the human population in check (Sieferle 2001; Merchant 2005). The escalation in the use of fossil fuels began when Europe's growing population was severely straining wood supplies and the average number of hours worked per person reached its historical peak in England, where the fossil fuel transition originated (Merchant 1980; Sieferle 2001; Pomeranz 2002). Fossil fuels allowed advances in technical and labor productivity that delayed Malthusian population problems—indeed, they triggered a population explosion (Stone 2001; Krausmann et al. 2008; Turner and Fischer-Kowalski 2010). Boserup's theory that, contrary to the Malthusian hypothesis, rising human population drives technological and other innovations to maintain adequate food production inadequately accounts for the temporary boost that fossil fuels injected into agricultural systems (Turner and Fischer-Kowalski 2010). The Green Revolution from the 1960s to 1990, still driven by fossil fuels, significantly increased agricultural yields and avoided renewed predictions of widespread famine that arose in the late 1960s (Khush 1999; McNeill 2000). However, current grain yield and production trends, along with the unlikelihood of compensating yield stabilization or decline with significant increases in agricultural land area, plus competition from biofuels, are giving rise to new warnings of increased famine in the twenty-first century (Khush 1999; Meadows et al. 2004; Smil 2007; Schade and Pimentel 2010). Malthus's predictions regarding population growth may have been prematurely right (Georgescu-Roegen 1971; Keyfitz 1991).

Declining global average fertility rates are slowing population growth, perhaps toward stabilization and decline in total numbers by the end of this century (United Nations 2010; Lee 2011; O'Neill et al. 2010). High birth rates in the 1950s and 1960s, longevity and age distributions, especially high proportions of young people, affect how much influence declining fertility rates will have on the total population over time (United Nations 2011). The momentum of demographic trends is such that even radical reductions in fertility worldwide, through coercion, massive short-term catastrophes (e.g. global nuclear war) or other means, hold little promise as quick fixes to concerns about overpopulation (Bradshaw and Brooks 2014). Wealth distribution, the level of education and reproductive freedom of women, the economics of supporting more old people, the migration of people to places with higher per capita consumption rates and the tendency for wealthy industrialized countries to invest capital in and reap profits from exploitation of natural resources and labor in poorer countries also influence population dynamics (Keyfitz 1991; Merchant 2005; United Nations 2011). Overall, under current trends it appears

> inevitable that the virtually locked-in increase in the global human population during the 21st century—regardless of trends in per capita consumption rates—risks increasing the threat to the environment posed by humans because of growing aggregate and accumulated demands.
>
> (Bradshaw and Brook 2014: 16615)

Views on the Earth's carrying capacity for humans are diverse and divergent (Keyfitz 1991; Daily and Ehrlich 1992; Meadows et al. 2004; Merchant 2005). The size of the human population does not necessarily correlate arithmetically with the extent of ecological impacts; for example, the rate of emissions of carbon dioxide in the past century was greater than the population growth rate (McNeill 2000), and the per capita rate of material use also grew (Krausmann et al. 2008). Yet, humans now appropriate between 25 and 40 percent of total net primary production (Vitousek et al. 1986; Haberl et al. 2007), and despite the complexities in how people ecologically affect Earth systems, human population is an important factor in the human-Earth relationship.

The increasing dispersal and mixing of people over time has had impacts that transcend the growth in human numbers. By no later than 13,000 years ago, humans had established a permanent presence in Africa, Europe, Asia, Oceania and the Americas (Ponting 2007). This distribution of the human population was very gradual, even if regional impacts on species distribution and land use could be significant. Beginning around 1500, long-range exploration, conquest and colonialism took hold, grounded in a need to find remote sources to provision the demands for food and other necessities in some regions, primarily Europe. In the modern globalized economy, the permanent or temporary movement of people for work, leisure or escape from danger or hardship is on another scale altogether, contributing to climate change and a host of other impacts.

References

Armelagos, G.J., A.H. Goodman and K.H. Jacobs. 1991. The origins of agriculture: population growth during a period of declining health. *Population and Environment* 13(1): 9–22.

Beddoe, R., R. Costanza, J. Farley, E. Garza, J. Kent, I. Kubiszewski, L. Martinez, T. McCowen, K. Murphy, N. Myers, Z. Ogden, K. Stapleton and J. Woodward. 2009. Overcoming systemic roadblocks to sustainability: the evolutionary redesign of world-views, institutions, and technologies. *Proceedings of the National Academy of Science* 106(8): 2483–2489.

Bradshaw, C.J.A., and B.W. Brooks. 2014. Human population reduction is not a quick fix for environmental problems. *Proceedings of the National Academy of Science* 111(46): 16610–16615.

Brown, P.G., and G. Garver. 2009. *Right Relationship: Building a Whole Earth Economy*. San Francisco: Berrett-Koehler.

Burke, E. III, and K. Pomeranz, eds. 2009. *The Environment and World History*. Los Angeles: University of California Press.

Carpenter, S.R., and E.M. Bennett. 2011. Reconsideration of the planetary boundary for phosphorus. *Environmental Research Letters* 6: 1–12.

Clark, B., and J.B. Foster. 2009. Ecological imperialism and the global metabolic rift: unequal exchange and the guano/nitrates trade. *International Journal of Comparative Sociology* 50(30–34): 311–334.

Cronon, W. 1993. The uses of environmental history. *Environmental History Review* 17(3): 1–22.

Crosby, A.W. 1987. The Columbian Voyages, the Columbian Exchange, and Their Historians. *Essays on Global and Comparative History*. Washington: American Historical Association.

Daily, G.C., and P.R. Ehrlich. 1992. Population, sustainability, and Earth's carrying capacity. *BioScience* 42(10): 761–771.

Davies, G. 2002. *A History of Money from Ancient Times to the Present Day*. Cardiff: University of Wales Press.

Diamond, J. 2005. *Collapse: How Societies Choose to Fail or Succeed*. New York: Viking Press.

Dow, G.K., C.G. Reed and N. Olewiler. 2009. Climate reversals and the transition to agriculture. *Journal of Economic Growth* 14: 27–53.

Ehrlich, P.R., and J.P. Holdren. 1972 (May). Critique of *The Closing Circle*. *Bulletin of the Atomic Scientists* 16–27.

Fischer-Kowalski, M., and H. Haberl. 1997. Tons, joules, and money: modes of production and their sustainability problems. *Society & Natural Resources* 10(1): 61–85.

Foley, J.A., N. Ramankutty, K.A. Brauman, E.S. Cassidy, J.S. Gerber, M. Johnston, N.D. Mueller, C. O'Connell, D.K. Ray, P.C. West, C. Balzer, E.M. Bennett, S.R. Carpenter, J. Hill, C. Monfreda, S. Polasky, J. Rockström, J. Sheehan, S. Siebert, D. Tilman and D.P.M. Zaks. 2011. Solutions for a cultivated planet. *Nature* 478: 337–342.

Foley, S.F., D. Gronenborn, M.O. Andreae, J.W. Kadereit, J. Esper, D. Scholz, U. Pöschl, D.E. Jacob, B.R. Schöne, R. Schreg, A. Vött, D. Jordan, J. Lelieveld, C.G. Weller, K.W. Alt, S. Gaudzinski-Windheuser, K-C. Bruhn, H. Tost, F. Sirocko and P. Crutzen. 2013. The Palaeoanthropocene—the beginnings of anthropogenic environmental change.

Garver, G. 2019. Confronting remote ownership problems with ecological law. *Vermont Law Review* 43(3): 425–454.

Georgescu-Roegen, N. 1971. *The Entropy Law and Economic Process*. Cambridge MA: Harvard University Press.

Giljum, S., A. Behrens, F. Hinterberger, C. Lutz and B. Meyer. 2008. Modelling scenarios towards a sustainable use of natural resources in Europe. *Environmental Science & Policy* 11: 204–216.

Haberl, H., K-H. Erb, F. Krausmann, V. Gaube, A. Bondeau, C. Plutzar, S. Gingrich, W. Lucht and M. Fischer-Kowalski. 2007. Quantifying and mapping the human appropriation of net primary production in Earth's terrestrial ecosystems. *Proceedings of the National Academy of Science* 104(31): 12942–12947.

Haberl, H., V. Gaube, R. Diaz-Delgado, K, Krauze, A. Neuner, J. Peterseil, C. Plutzar, S.J. Singh and A. Vadineanu. 2009. Towards an integrated model of socioeconomic biodiversity drivers, pressures and impacts. A feasability study based on three European long-term socio-ecological research platforms. *Ecological Economics* 68: 1797–1812.

Hall, C.A.S., and K.A. Klitgaard. 2012. *Energy and the Wealth of Nations: Understanding the Biophysical Economy.* New York: Springer.

Harari, Y.N. 2014. *Sapiens: A Brief History of Humankind.* Toronto: McClelland & Stewart.

Hoffman, R.C. 2001. Frontier foods for late medieval consumers: culture, economy, ecology. *Environment and History* 7: 131–167.

Jacobsen, T., and R.M. Adams. 1958. Salt and silt in Ancient Mesopotamian agriculture. *Science* 128(3334): 1251–1258.

Johnson, N. 2006. Swine of the times: the making of the modern pig. *Harper's Magazine* (May): 47–56.

Jones, E.L. 2003. *The European Miracle: Environments, economies and geopolitics in the history of Europe and Asia, 3rd edition.* New York: Cambridge University Press.

Kallis, G. 2018. *Degrowth.* Newcastle upon Tyne, UK: Agenda Publishing.

Keyfitz, N. 1991. Population and development within the ecosphere: one view of the literature. *Population Index* 57(1): 5–22.

Khush, G.S. 1999. Green revolution: preparing for the 21st century. *Genome* 42: 646–655.

Krausmann, F., M. Fischer-Kowalski, H. Schandl and N. Eisenmenger. 2008. The global sociometabolic transition. *Journal of Industrial Ecology* 12(5/6): 637–656.

Krausmann, F., S. Gingrich, N. Eisenmenger, K.H. Erb, H. Haberl and M. Fischer-Kowalski. 2009. Growth in global materials use, GDP and population during the 20th century. *Ecological Economics* 68: 2696–2705.

Lee, R. 2011. The outlook for population growth. *Science* 333: 569–573.

McNeill, J.R. 2000. *Something New Under the Sun: An Environmental History of the Twentieth-Century World.* New York: W.W. Norton & Co.

McNeill, J.R. 2003. Observations on the nature and culture of environmental history. *History and Theory* 42(4): 5–43.

McNeill, J.R. 2010a. The state of the field of environmental history. *Annual Review of Environment and Resources* 35: 345–374.

McNeill, J.R. 2010b. *Mosquito Empires: Ecology and War in the Greater Caribbean.* New York: Cambridge University Press.

Meadows, D., J. Randers and D. Meadows. 2004. *Limits to Growth: The 30-Year Update.* White River Junction, VT: Chelsea Green Publishing Company.

Merchant, C. 1980. *The Death of Nature: Women, Ecology and the Scientific Revolution.* San Francisco: Harper & Row.

Merchant, C. 2005. *Radical Ecology: The Search for a Livable World.* New York: Routledge.

Moore, J.W. 2007. Silver, Ecology, and the Origins of Modern World, 1450–1640, pp. 123–142 in A. Hornborg, J.R. McNeill and J. Martinez-Alier, eds., *Rethinking Environmental History: World-system History and Global Environmental Change.* Toronto: AltaMira Press.

Nyquist, B., A. Persson, F. Moberg, L. Persson, S. Cornell and J. Rockström. 2013. *National Environmental Performance on Planetary Boundaries: A study for the Swedish Environmental Protection Agency.* Stockholm: Swedish Environmental Protection Agency.

Oldfield, F., A.D. Barnosky, J. Dearing, M. Fischer-Kowalski, J. McNeill, W. Steffen and J. Zalasiewicz. 2014. *The Anthropocene Review:* Its significance, implications and the rationale for a new transdisciplinary journal. *The Anthropocene Review* 1(1): 3–7.

Olshansky, S.J., B.A. Carnes and A. Désesquelles. 2001. Prospects for human longevity. *Science* 291(5508): 1491–1492.

O'Neill, B.C, M. Dalton, R. Fuchs, L. Jiang, S. Pachauri and K. Zigova. 2010. Global demographic trends and future carbon emissions. *Proceedings of the National Academy of Science* 107(41): 17521–17526.

Pearce, F. 2012. *The Land Grabbers: The New Fight Over Who Owns the Earth.* Boston: Beacon Press.

Pomeranz, K. 2002. Political economy and ecology on the eve of industrialization: Europe, China, and the global conjuncture. *The American Historical Review* 17(2): 425–446.

Pomeranz, K. 2009. Introduction: World History and Environmental History, pp. 3–32 in E. Burke III and K. Pomeranz, eds., *The Environment and World History.* Los Angeles: University of California Press.

Ponting, C. 2007. *A New Green History of the World: The Environment and the Collapse of Great Civilizations.* Toronto: Penguin Books.

Pryor, F.L. 1986. The adoption of agriculture: some theoretical and empirical evidence. *American Anthropologist* 88: 879–894.

Ramankutty, N., A.T. Evan, C. Monfreda and J.A. Foley. 2008. Farming the planet: 1. Geographic distribution of global agricultural lands in the year 2000. *Global Biogeochemical Cycles* 22: GB1003.

Rengasamy, P. 2006. World salinization with emphasis on Australia. *Journal of Experimental Botany* 57(5): 1017–1023.

Richards, J.F. 2003. *The Unending Frontier: An Environmental History of the Early Modern World.* Berkeley CA: University of California Press.

Robock. 2008. 20 reasons why geoengineering may be a bad idea. *Bulletin of the Atomic Scientists* 64(2): 14–18.

Rockström, J., W. Steffen, K. Noone, Å. Persson, F.S. Chapin III, E. Lambin, T.M. Lenton, M. Scheffer, C. Folke, H. Schellnhuber, B. Nykvist, C.A. De Wit, T. Hughes, S. van der Leeuw, H. Rodhe, S. Sörlin, P.K. Snyder, R. Costanza, U. Svedin, M. Falkenmark, L. Karlberg, R.W. Corell, V.J. Fabry, J. Hansen, B. Walker, D. Liverman, K. Richardson, P. Crutzen and J. Foley. 2009. Planetary boundaries: exploring the safe operating space for humanity. *Ecology and Society* 14(2): 32 [online] URL: www.ecologyandsociety.org/vol.14/iss2/art32/

Schade, C., and D. Pimentel. 2010. Population crash: prospects for famine in the twenty-first century. *Environment, Development & Sustainability* 12: 245–262.

Scott, J.C. 1998. *Seeing Like a State.* New Haven CT: Yale University Press.

Sieferle, R.P. 2001. *The Subterranean Forest: Energy Systems and the Industrial Revolution.* Cambridge UK: The White Horse Press.

Smil, V. 2007. *Energy in Nature and Society.* Cambridge, MA: The MIT Press.

Sörlin, S., and P. Warde. 2007. The problem of the problem of environmental history: a re-reading of the field. *Environmental History* 12: 107–130.

Steffen, W., J. Grinevald, P. Crutzen and J. McNeill. 2011. The Anthropocene: conceptual and historical perspectives. *Philosophical Transactions of the Royal Society A* 369: 842–867.

Steffen, W., K. Richardson, J Rockström, S.E. Cornell, I. Fetzer, E.M. Bennett, R. Biggs, S.R. Carpenter, W. de Vries, C.A. de Wit, C. Folke, D. Gerten, J. Heinke, G.M. Mace, L.M. Persson, V. Ramanathan, B. Reyers and S. Sörlin. 2015. Planetary boundaries: guiding human development on a changing planet. *Science* 347(6223): 1259885-1–1259885-10.

Stone, G.D. 2001. Theory of the square chicken: advances in agricultural intensification theory. *Asia Pacific Viewpoint* 42(2/3): 163–180.

Testart, A. 1982. The significance of food storage among hunter-gatherers: residence patterns, population densities, and social inequalities. *Current Anthropology* 23(5): 523–537.

Turner, B.L. II and M. Fischer-Kowalski. 2010. Ester Boserup: an interdisciplinary visionary relevant for sustainability. *Proceedings of the National Academy of Science* 107(51): 21963–21965.

United Nations. 1999. *The World at Six Billion.* New York: Department of Economic and Social Affairs, Population Division.

United Nations. 2010. *Millennium Development Goals Report 2010.* New York: Department of Economic and Social Affairs.

United Nations. 2011. *World Population Prospects, the 2010 Revision.* New York: Department of Social and Economic Affairs, Population Division.

United Nations. 2019. *World Population Prospects: Highlights.* New York: United Nations.

United Nations Environment Programme (UNEP). 2011. *Towards a Green Economy: Pathways to Sustainable Development and Poverty Eradication—A Synthesis for Policymakers.* Nairobi: UNEP. [Online]: www.unep.org/greeneconomy/GreenEconomyReport/tabid/29846/language/en-US/Default.aspx

Vitousek, P.M., P.R. Ehrlich, A.H. Ehrlich and P.A. Matson, 1986. Human appropriation of the products of photosynthesis. *BioScience* 36(6): 368–373.

4 How conceptual factors like law helped lead to the Anthropocene

Legal systems are key conceptual constructs—that is, socially organized intellectual narratives, frameworks or structures—that have both emerged from and helped to shape the trajectory of humanity towards the Anthropocene. Legal systems are inevitably interlinked with other ecological and socially constructed systems in the global panarchy. They are largely instrumental, in service of the underlying social, economic and political norms and narratives that prevail (M'Gonigle and Takeda 2013).

Definitional issues regarding law and legal systems

Defining law is always challenging, in part because distinctions among rules of conduct derived from domains such as tradition, religion, ethics and encoded law are invariably evolving and difficult to discern (Ross 1966). Broadly, law "is the concrete expression of the practical necessities of community life" (Prince 1904: 606), embedded in society through a system of rules, institutions and mechanisms that "bind, constrain, regularise and civilise" (Cullinan 2011: 29). Law sorts out social understandings of legally enforceable entitlements (or rights), freedoms, obligations and prohibitions (Ruhl 1996) that apply to people and other entities treated as legal persons or rights holders. The enforceability of law in a society distinguishes legal imperatives from moral ones, which can have varying degrees of weight in ordering human societies. In a pluralistic view of law, the range of mechanisms—whether formal, cultural, customary or otherwise influential—that operate effectively to regulate behavior more or less predictably within a communal context can be quite broad (M'Gonigle 2008; M'Gonigle and Takeda 2013).

Theories of natural law and of legal positivism are at opposite ends of a continuum of legal philosophy regarding the nature of law and its connection to moral foundations. Natural law is founded on universal and unalterable moral notions of right and wrong; the laws that a society adopts derive from those moral underpinnings, taking into account practical notions of collective life (Ross 1966; Ponting 2007; Cullinan 2011). The moral foundation of natural law, whether based on religious or secular conceptions of right and wrong, is by definition beyond dispute, and thus invincible to argument based on empirical evidence (Ross 1966; Sen 1999; Ponting 2007; Cullinan 2011). By contrast, legal positivism—or, in American legal theory, legal formalism (Ruhl 1996)—rejects any notion that laws are, or should be, tied rationally to a fixed, supreme moral foundation. To the legal positivist, the law is simply what the lawmaker says it is, and references to morality or reason are irrelevant.

Natural law derives its legitimacy by insisting that its moral foundations are beyond dispute. Yet, opposing ideologies will inevitably arise as to what those universal moral foundations are (Elliott 2005; Ponting 2007). Because it takes the ultimate truth of those foundations to be axiomatic, natural law contains no space in which to resolve opposing convictions regarding them. To some, inequality among humans has been ordained as natural and good, and therefore slavery and other forms of domination and social hierarchy have been justified with reference to a solid moral foundation. To others, slavery is near the apex of immorality. To the writers of the United States Constitution and the intellectual forces behind the French Revolution, the right to own property was rooted in natural law (Ost 2003; Nadeau 2006; Capra and Mattei 2015). According to other worldviews, individual property rights are an affront to the interconnectedness of all life, and the inter-subjectivity of all living things (Berry 1999; Cullinan 2011; Burdon 2015). Further, empirical evidence can emerge that calls into question presumably immutable moral precepts or that reveals irreconcilable conflicts among them (Elliott 2005).

At the extreme, legal positivists' rejection of any rational connection of law to a moral foundation is unhinged from reality. A persuasive critique of legal positivism is that it envisions a static and reductionist set of rules that ignore the dynamic nature of law and its interactive temporal and spatial connection to the evolving social and ecological contexts in which it is embedded (Ruhl 1996; Burdon 2015; Capra and Mattei 2015). Legal positivism implicitly presupposes a degree of predictability and certainty at odds with law's own dynamic qualities and the dynamic qualities of the contexts to which it applies.

Legal realism emerged as a possible way to resolve the impossibilities of both natural law's implicit rejection of evolving, empirically informed notions of ethics and morality, and also of legal positivism's blindness to moral context. Legal realism sought to reconnect law empirically with science, but it retained an insistence on predictability and certainty, treating scientific information as static and determinative (Ruhl 1996). An emerging notion of law and legal systems as complex adaptive systems that exhibit emergence, non-linearity and other features of complex systems provides a more promising basis for analyzing the historical trajectory of law and imagining legal systems suited to the unprecedented challenges of the Anthropocene. This view takes into account the evolutionary nature of the interlinked socially constructed systems and ecological systems that make up the global panarchy, including core moral foundations and worldviews that underlie legal systems and other socially constructed systems.

As used in this book, law and legal systems are dynamic systems that inevitably evolve, along with worldviews that emerge from and shape interrelated social, economic, political and cultural systems. Although law expresses the "practical necessities of community life" (Prince 1904: 606), the conception of those necessities evolves. Law therefore builds on itself, such that "every law-code in existence ... is a combination ... of certain ancient, fixed principles descending from the days of the first nationalization of the people who evolved the code, and ... of various later modifications which became necessary as the national life ripened under the influence of new conditions, incidental to a growing culture" (Prince 1904: 606).

Overview of the historical development of law and legal systems

Considering how and why early forms of law emerged helps elucidate core elements of legal systems generally. What were the conditions that first gave rise to enforceable, socially embedded rules that reflected desired norms of behavior within a defined human community? How has the need for communal acceptance of a set of rules arisen and evolved in different contexts?

Codified law first emerged in the Mesopotamian civilizations that took hold as agriculture spread in the region served by the Tigris and Euphrates rivers beginning around 6,000 years ago. The development of agriculture led to a steady concentration of a growing population, and consequently to increasing stratification of the roles and status of people in society (Testart 1982; Davies 2002; Ponting 2007; Burdon 2015). Most significantly, agriculture led to the emergence of an elite class that relied on surplus production of food from farmers (Ponting 2007). Across settled societies, this elite sector comprised about five to ten percent of the population during the long reign of pre-industrial societies (Ponting 2007). This stratification of society gave way inherently to a differentiated understanding of the entitlements, freedoms, obligations and prohibitions that applied to different people—an undercurrent that persists in modern societies with, for example, ongoing differentiations based on race and gender (Burdon 2015). As well, it gave rise to the need to justify this differentiation—most often by aligning social hierarchies with military authority or systems of religious belief (Ponting 2007).

Two of the first recognized legal codes are the Code of Ur-Nammu from about 2250 BC and the later Code of Hammurabi, but legal codes may have first arisen around 5,000 years ago (Prince 1904; Kornfeld 2009). Early legal codes were somewhat random enumerations of rules of social order, as opposed to the structured and more comprehensive modern legal codes. The Code of Hammurabi was grounded in a notion of divine supremacy that established notions of right and wrong and social order, with the ruler serving as the agent and ultimate judge of divine authority (Prince 1904; Kornfeld 2009). The Hammurabi code set down enduring notions of "an eye for an eye, a tooth for a tooth," as well as rules, definitions and punishments for such things as adultery, divorce, false witness, rape, theft, commercial fraud and negligence (Prince 1904). It also implicitly recognized and codified the class structure that arose with the establishment of urban centers (Ponting 2007). Upper class "householders," the free "poor" classes, and slaves—each with subdivisions (court officials, priests, tailors, shepherds, etc.) that took shape as specialization emerged—all had different legal status in the code (Prince 1904).

The relationship between law and the impacts of societies on their ecological foundations also took hold in this era. Ancient Mesopotamian law included provisions related to water use, such as rules that established liability for those injured by failure to maintain proper irrigation systems (Kornfeld 2009). Water and irrigation systems were treated as communal rather than individual assets (Kornfeld 2009). Nonetheless, by mediating rights and responsibilities of individual users of the shared resource, ancient water law inherently recognized notions of individual entitlements to property.

> [A] rich legal system, which grew out of an urbane and highly developed social and political system, was necessary to produce and maintain elaborate and complex irrigation works. In addition, a well-developed water regulatory system protected both a collective or individual ownership of the resources.
>
> (Kornfeld 2009: 34)

However, even then, those laws were inadequate to prevent the creeping (and perhaps unrecognized) salinization of Mesopotamian irrigation waters that eventually caused a sharp decline in agricultural output and a consequent decline of Mesopotamian civilizations (Rengasamy 2006; Ponting 2007; Kornfeld 2009).

Ever since these origins, law has tended to reflect the belief systems and accordant power structures in which it has emerged and evolved. In many well-documented cases of societal collapse, as with the gradual and catastrophic salinization of Mesopotamian irrigation water, some kind of legal order was no doubt implicit in the socio-political trajectory to which broad communities of people bound themselves as they headed toward disaster, unaware or dismissive of slow-moving dangers (Diamond 2005; Beddoe et al. 2009). However, legal systems of various kinds also allowed expansionist or developmentalist projects to succeed, and civilizations to prosper and spread.

In China, complex societies consisting of large urban centers, industrial systems of production and extensive trade networks, all contributing to expansion, existed as early as the eleventh century (Ponting 2007). These complex social systems could not function without a fairly sophisticated legal order. Parallel to the expansion of East Asian cultures and their legal regimes, Europeans in the early modern era began a significant and ongoing trend toward global interconnectedness through long-range travel and colonization. An expanded European-based legal order emerged. Generally, divine authority of monarchs provided the moral and legal authority to support the competitive quest by European countries for conquest and domination of new lands to support economies that European ecosystems could no longer sustain (Hoffman 2001; Ponting 2007).

European feudal systems by and large controlled the use of land and resources, and the law increasingly imposed constraints—for example, on exploitation of forest resources or hunting of fur-bearing animals—as the rebounding population following the Black Death of the fourteenth century increased environmental stresses (Sieferle 2001; Ponting 2007). Privateers acting on behalf of European monarchs through official charters in that period evolved into the first corporations—essentially, legally recognized artificial entities that allowed investors to reap vast material returns from conquest of new territory and resources in the Americas, Africa and Asia (Korten 2006). European monarchs sometimes relied on papal bulls that they interpreted as granting them permission to assert sovereignty over "discovered" non-Christian lands with a view to converting any inhabitants to Christianity (*Romanus Pontifex* 1455, *Inter Caetera* 1493). The imprimatur of divinely rooted royal right infused the violent European invasions and conquests of the early modern era with moral and legal authority that implicitly persists in the modern societies that emerged in conquered lands in the Americas and elsewhere —in the sense that the legitimacy of those colonial conquests is rarely comprehensively or effectively questioned or redressed on the basis of contemporary notions of the rule of law, human rights and justice (M'Gonigle 2008). This is true not only in regard to those societies domestically, but also in regard to global patterns of material and energy flows through trade channels that have roots in the power imbalances and legal orders of the early modern era (Martinez-Alier 1987; Ponting 2007; Escobar 2010).

In Europe, the Enlightenment grew out of the new arrangements of power and wealth to which European expansionism and colonialism significantly contributed, yielding a new paradigm to which the legal order responded in kind. Through the philosophies of key figures such as René Descartes and Francis Bacon, the Enlightenment involved both a significant step in detaching humans from nature and an incursion of scientific reasoning and method on the domain of divine authority (Merchant 1980; Ost 2003; Capra and Mattei 2015). In conjunction with this transition, the Peace of Westphalia in 1648 provided a blueprint for more peaceful relations among sovereign nations. However, the resulting focus on sovereignty also yielded a troubling foundation for the current selective invocation of sovereignty to impede progress toward an effective supranational legal structure to address many critical transboundary or global ecological issues (Brown and Garver 2009; Garver 2015). The role of State sovereignty as a key conceptual factor is discussed further below. The modern State-centered international legal order arose from the Westphalian accord.

Parallel to the European Enlightenment, a moral philosophy emphasizing individual freedoms, unencumbered markets and private property rights arose, challenging the power of monarchies and ultimately leading to the American and French Revolutions in the eighteenth century and an overall decline of royal powers in Europe (Ost 2003). Different ideological orientations have undergirded legal systems that supported developmentalist regimes in other parts of the world leading to the current era (Pomeranz 2009). Nonetheless, the Western model of individual freedoms and equal opportunity, State-centered capitalist economies, subjugation and exploitation of ecosystems for human ends and, ultimately, a core commitment to sustained economic growth has become dominant in the increasingly globalized modern world.

The role of law and related conceptual factors in the current human dilemma

The modern global legal system reflects a hierarchy of values by which nations collectively give priority to powerful and wealthy economic actors seeking as much profit as quickly as possible while satisfying demands of a growing global consumer culture for employment, goods and services (Capra and Mattei 2015). These national priorities are typically founded on sovereign interests, protection of competitive advantage for domestic enterprises, private property rights and an overriding commitment to sustained economic growth. This hierarchy of values reflects power dynamics that infuse the global infrastructure for socially constructed normative systems—economic, political, social, cultural—that underlie and interact with the legal system. Within this hierarchy, aspirational commitments to intergenerational, intragenerational and interspecies fairness and justice and to an enduring harmonious human-Earth relationship emerged (U.S. National Environmental Policy Act 1970; United Nations 1987). These typically soft commitments (Brooks et al. 2002) might be seen as counterbalancing the idealistic assumption that free markets produce fair and just outcomes and adjust themselves through substitution and innovation to declines in the deterioration of the ability of ecological sources and sinks of the throughput of material and energy to support the economy. However, the primacy of growth insistence that prevails in the current ordering of human-Earth relations (Jackson 2009) has routinely subjugated those vague aspirational notions. The following sections briefly explain factors that contribute to this hierarchy of values within the legal system.

Overarching commitment to sustained economic growth

The globally pervasive capitalist fixation on growth is a twentieth-century phenomenon that solidified after the Great Depression (McNeill 2000; Speth 2008), but its historical antecedents go back centuries. "Developmentalism" describes growth-oriented intellectual framings encompassing not only capitalism, but also the expansionist, growth-oriented projects built on other ideologies, such as in Mughal India, both Czarist Russia and the Soviet Bloc and China during the Ming and Qing dynasties and Mao's failed Great Leap Forward (Scott 1998; McNeill 2000; McNeill 2003; Richards 2003; Burke and Pomeranz 2009). The emergence in neo-classical economics of an insistence on growth based on capitalism is a special case. The essence of capitalism is privately owned and financed production of goods and services for profit, where key decisions are made by profit-driven individuals or corporations (through shareholders and directors) based on market-based criteria. Capitalism warrants singling out because, with the collapse of the Soviet Union and China's embrace of market capitalism, it has emerged as the dominant global form of the "state religion" of economic growth that has existed in different political systems across history (McNeill 2000; Pomeranz 2002). Moreover, capitalism most clearly favors forms of economic expansion that aim to maximize profits, with the least practicable consideration of other societal goals (Moore 2003).

The dominant societal objective internationally currently is to raise and perpetually maintain economic growth. The leaders of the Group of Twenty (G20) nations in 2019 committed "to foster global economic growth, while harnessing the power of technological innovation, in particular digitalization, and its application for the benefit of all" and "to strive to create a virtuous cycle of growth by addressing inequalities and realize a society where all individuals can make use of their full potential" (G20 2019). Their declaration reaffirmed numerous earlier commitments to sustained economic growth (G20 2010; G20 2011; G20 2013; G20 2014; G20 2015). The G20's declarations are consistent with the international community's prevailing conception of sustainable development as development "that meets the needs of the present without compromising the ability of future generations to meet their own needs" (United Nations 1987: Ch. 2, para. 1). The principles set out in the 1992 Rio Declaration on Environment and Development indicate a broad international commitment to environmental protection, but the sustainable development they connote requires an overriding and intractable commitment to economic growth.

The "green economy" strategy of the United Nations Environment Programme (UNEP 2011) maintains this growth-insistent interpretation of sustainability. Greening of the global economy is cast as a "new engine of growth" that builds, rather than runs down, social and natural capital (UNEP 2011: 16). In this vision of sustainability, decoupling sustained economic growth from ecological impacts is fundamental (UNEP 2011). As well, the outcome document of the Rio+20 Summit in 2012, *The future we want*, refers to nations' shared commitment to economic growth, or "sustained, inclusive and equitable economic growth" ("sustainable" has morphed into "sustained") more than 20 times. The Sustainable Development Goals (SDGs) that the United Nations adopted in 2015 are likewise growth-insistent (United Nations 2015a). The SDGs include a range of environmental, social and economic agendas up to 2030, including commitment to "promote sustained, inclusive and sustainable growth" in SDG 8 (United Nations 2015a, para 13).

The world economy was growing both in terms of its throughput of material and energy and in terms of the aggregate monetary value of its goods and services before economic growth solidified as the overarching macroeconomic objective for economies around the world, from the global to the local (McNeill 2000). A conscious conceptual orientation toward economic growth can mean material growth, or growth in purely economic terms (gross domestic product, GDP, or gross national product, GNP) (Daly 1996). Resource productivity and dematerialization, historically tempered by the rebound effect (or Jevons paradox) by which gains in efficiency lead to growth in overall consumption, determine the extent of the correlation between material growth and GDP growth

(Daly 1996; Daly and Farley 2004; Jackson 2009; Krausmann et al. 2009; von Weizsäcker et al. 2009). Trust in the emergence of technological solutions that decouple economic growth from the throughput of material and energy in human society and its impacts is the only way to maintain growth in this growth-insistent paradigm on a finite planet (Beddoe et al. 2009; Jackson 2009; M'Gonigle and Takeda 2013; Kallis 2018). Yet, insisting on either type of growth is problematic, first because of ecological limits on the Earth's capacity to provide material and energy inputs and absorb their outputs, and second because of physical limits on the possibilities for dematerialization (Daly 1996; Daly and Farley 2004; Jackson 2009; Kallis 2018).

Capitalism is a particularly virulent means to maintain the global fixation on growth in GDP as a necessary element of socio-economic policy. A powerful set of socio-political forces support growth insistence under a capitalist rubric.

> An unquestioning society-wide commitment to economic growth at almost any cost; enormous investment in technologies designed with little regard for the environment; powerful corporate interests whose overriding objective is to grow by generating profit, including profit from avoiding environmental costs they create; markets that systematically fail to recognize environmental costs unless corrected by government; government that is subservient to corporate interests and the growth imperative; rampant consumerism spurred by a worshipping of novelty and by sophisticated advertising; economic activity so large that its impacts alter the fundamental biophysical operations of the planet—all combine to deliver an ever-growing world economy that is undermining the planet's ability to sustain life.
>
> (Speth 2008: 7–8)

The liberal market economy of capitalism has become disembedded from social contexts that formerly had the effect of tempering the profit motive with concern for the common good (Polanyi 1957). In modern capitalism, the increasingly unregulated ability to manipulate the financial system and engage in financial speculation produces economically, socially and ecologically disruptive short-term capital bubbles and vast amounts of virtual wealth untied from real assets or local contexts, including the ecological base of the economy (Martinez-Alier 1987; Daly and Farley 2004; Pearce 2012; Klein 2014).

A few stark examples illustrate the roots of the capitalist project of expansion and growth, starting with the history of medieval and early modern mining in Europe and the Americas. Capitalism arose in Europe at a time when feudal systems were giving way to private property systems that are an integral part of capitalism's evolution (Richards 2003; Richards 2009). Socially and ecologically disruptive silver and other metal mining in central Europe in the fifteenth century, which later expanded to the Americas, involved "huge capital investment, large-scale industrial production, and aggressive monetization" (Moore 2007: 123). In the early modern period, European capitalists, with the Dutch

East India Company—a precursor to the modern multinational corporation—a leading example, financed continually expanding commercial ventures in Europe, Africa and Asia, with far-reaching social and environmental implications, and massive accumulation of wealth and a rise in the political power of capitalist elites in Western Europe (Richards 2003; Korten 2006).

A more recent example is the rapacious, British-led, ecologically imperialistic rush on guano and nitrate deposits in Peru and Chile in the nineteenth century (Clark and Foster 2009). Guano and nitrates, extracted under harsh labor conditions by indentured Chinese workers, yielded handsome profits for foreign investors and wealthy Peruvian and Chilean elites, but exploitative trade arrangements, diminishing stocks and inorganic fertilizers ultimately left the Peruvian and Chilean economies in debt and in tatters (Clark and Foster 2009). A sequel to this example of "ecologically unequal exchange" (Clark and Foster 2009) is the case of Nauru, a formerly self-sufficient Pacific island that in 1900 was found to contain high quality phosphates (Gowdy and McDaniel 1999). A century later, the phosphate has been exhausted, with most of the proceeds siphoned off to foreign investors. The island's ecology and socio-cultural structure have been upended, and the people of Nauru now depend on foreign aid and the proceeds from a trust fund that was built up with a small percentage of the profits of phosphate mining but decimated by bad management and world market upheavals (Gowdy and McDaniel 1999). Exhaustion of phosphates on Nauru illustrates the shaky foundations of the weak sustainability version of capitalism, by which all forms of capital—economic, social or natural, for example—can be expressed in monetary terms and substituted for one another without limit (Gowdy and McDaniel 1999; Speth 2008).

The globalized market that is the modern version of the world market that arose from Chinese and European trading in the early modern period (Crosby 1987; McNeill 2000; Richards 2003) operates on the principles of capitalism featured in the foregoing examples. Capital now moves with few constraints around the globe, always seeking to expand and maximize short-term profit, engaging countries and multinational corporations in a frenzy of market competition that relies on the greatest exploitation of cheap labor and lowest-cost extraction of material and energy possible (McNeill 2000; Rees 2002; Speth 2008; Pearce 2012). In all the foregoing examples, as with the recent rash of land and resource grabs, particularly in Africa, Latin America and Asia, and bioprospecting in tropical forests, the owners and controllers of financial capital, along with political leaders they support or tolerate, retain authority and control over the flows of material and energy (Clark and Foster 2009; Pearce 2012; Garver 2019).

The political power of wealthy corporations and individuals is maintained with massive amounts of commercial advertising and public opinion campaigns (Speth 2008). A stark example is the well-financed, politically-connected and market fundamentalist scientists—the "merchants of doubt"—who aggressively pursued the "Tobacco Strategy" of swaying American public opinion by raising doubts about the solid scientific consensus behind the cancer risks of smoking, human-caused depletion of stratospheric ozone, acid rain from coal-fired power plants and anthropogenic climate change (Oreskes and Conway 2010).

That capitalistic economic growth has become ideologically locked in around the world is a paradox, in that it has roots in the Enlightenment ideal of rationality but has now entered a global ecological context that imbues it with the irrationality of destroying one's home, or one's offspring and their future. In sum, "[c]enturies of economic achievement have produced the cultural expectation of an ever-expanding economy flowing seamlessly from its past" (M'Gonigle and Takeda 2013: 1012). The widespread commitment to growth and fear of reckoning with conventional macroeconomics' ecological blind spots has attained the kind of romantic idealism, feeling and passion that Enlightenment thinkers set about to master and control (Merchant 1980; Sörlin and Warde 2007).

Because true long-term sustainability cannot be growth-insistent, the current overriding commitment to economic growth and the subsidiary goal to raise living standards throughout the world to those in wealthy countries must eventually give way to a more hopeful alternative. A counternarrative is emerging that challenges insistence on infinite economic growth on a finite planet in the Anthropocene. This counternarrative, grounded in systems science and an understanding of how Earth's life systems behave and evolve, gives top priority to the systems impossibilities and long-term catastrophic socio-ecological consequences if the economy grows infinitely, and economic and political trade-offs continue to outweigh non-negotiable ecological limits. Under any definition of sustainability, collective action to reduce material and energy demands of human society is essential, with law playing a crucial role (Jackson 2009; UNEP 2011; M'Gonigle and Takeda 2013). However, economic growth has a strong historical correlation with ecological degradation and persistent social injustice, and at some point, sufficient decoupling of ecological degradation from growth becomes impossible (Daly 1996; Jackson 2009; Krausmann et al. 2009; McLellan et al. 2014; Kallis 2018). In Parts III and IV of this book, I analyze possible alternatives and potential pathways for pursuing an approach to law that would support them.

Reductionism and lack of holistic, systems-based approaches to law

The current globally dominant macroeconomic model lacks rigorous and scientifically realistic means to link overarching models of utility-driven market dynamics with ecosystem dynamics that are thermodynamically constrained (Georgescu-Roegen 1971; Daly 1996). The main omission that results from this misalignment is the failure to integrate into socially constructed normative systems, such as law and economics, systems-based responses to aggregate, multi-scalar and interconnected drivers and impacts of human activities (Capra and Mattei 2015). Legal systems and related normative social systems lack sufficient mechanisms to track and respond to aggregate impacts of human activities on planetary ecological boundaries of safe operating space for humanity (Garver 2013; Nyquist et al. 2013). According primal weight to aggregate ecological impacts in light of ecological limits is difficult because tracking the innumerable human drivers of the full panoply of environmental change related to planetary boundaries at and across different spatial and temporal scales, which are increasingly remote and interrelated, is immensely challenging and imbued with uncertainty (Nyquist et al. 2013).

Instead of following a holistic, systems-based approach, the United States legal system, as well as others, tends to treat the many long-term, cumulative impacts or seemingly minor impacts as speculative or inconsequential and therefore legally unimportant (Bugge 2013). The result is a reductionist, as opposed to holistic, tendency in environmental law to focus on environmental problems as temporally and spatially discrete and isolated (Boyd 2004; Capra and Mattei 2015). The reductionist approach in environmental law obscures consideration of the aggregate impacts of human activity on the global ecosphere and impedes the full integration of a systems-based ecological approach into the entire legal infrastructure (Ruhl 1996; Capra and Mattei 2015). Related to this, the currently dominant economic paradigm incorporates an implicit instrumentalist assumption that humans stand apart from, and reign over, the environment and non-human species (Berry 1999; Kysar 2002; Garver 2013).

Contemporary environmental law does include passing references to holistic approaches. For example, the rising environmental awareness in the United States in the 1960s and 1970s included growing concern about the finiteness of "spaceship Earth" (Boulding 1975; Brown and Garver 2009), an influential "deep ecology" movement (Naess 1984) and a greater focus on seeing humans as integrated in the Earth's biotic community (Leopold 1949; Delgado 1991; Tarlock 2000). Reflecting these concerns, the opening section of the United States *National Environmental Policy Act (NEPA)* states:

> The Congress, recognizing the profound impact of man's activity on the interrelations of all components of the natural environment, particularly the profound influences of population growth, high-density urbanization, industrial expansion, resource exploitation, and new and expanding technological advances and recognizing further *the critical importance of restoring and maintaining environmental quality to the overall welfare and development of man*, declares that it is the continuing policy of the Federal Government … to use all practicable means and measures … in a manner calculated to foster and promote the general welfare, *to create and maintain conditions under which man and nature can exist in productive harmony, and fulfill the social, economic, and other requirements of present and future generations of Americans*.[1]

However, hard ecological limits implied by deep ecology and related paradigms were too much "a radical break from the western philosophical and legal tradition" (Tarlock 2000: 240). Thus, environmental laws in the United States virtually never give ultimate primacy to ecological limits over economic objectives or technological limitations (Garver 2013). Moreover, in the United States, economically driven ideologies have increasingly crowded out the deep environmental conscientiousness of the 1960s and 1970s (Heinzerling 1995; Westra 2011). In Part II of this book, I present details of some of the failures of contemporary environmental law to adequately respect planetary ecological limits.

The reductionist approach that dominates legal systems that play a role in regulating the human-Earth relationship impedes full integration of ecological considerations in legal and policy infrastructures (Bosselmann 2008; Capra and Mattei 2015). Instead, environmental law tends to be isolated and secondary in legal systems, a boutique area of law (M'Gonigle and Takeda 2013) with a set of norms and procedures that lose ecological coherence, enforceability and weight in decision-making, the more environmental impacts of actions or decisions to which those norms or procedures apply are spatially and temporally remote, or significant only cumulatively. The Earth Charter (2000) expresses well the need to integrate an ecologically bounded notion of law throughout the legal system:

> To move forward we must recognize that in the midst of a magnificent diversity of cultures and life forms we are one human family and one Earth community with a common destiny. We must join together to bring forth a sustainable global society founded on respect for nature, universal human rights, economic justice, and a culture of peace. Towards this end, it is imperative that we, the peoples of Earth, declare our responsibility to one another, to the greater community of life, and to future generations.

Legal protection of sovereignty and State-centered globalization

The structure and functioning of the current international system of sovereign nation states supports growth-insistent national agendas and resists effective supranational governance regimes for containing human activities within ecological limits. Modern nation states are social constructs that evolved from civilizations with recognized rulers, hierarchies and specialized trades that arose following the Neolithic Transition (Ponting 2007; Dow et al. 2009). Throughout history, central authorities, and in time States, emerged with a view to establishing their boundaries of authority, managing property and other economic arrangements within their control and consolidating economic capital and wealth for developmentalist projects—often with a bent toward conquest and expansion (Richards 2003; Ponting 2007; Pomeranz 2009). The world is now entirely carved up geopolitically into territories claimed by nation states individually or in common (e.g. Antarctica) (Richards 2009).

Since the early modern era, States and other central authorities have interacted increasingly on an international and multinational level, through peaceful trade and other alliances as well as through invasion and armed conflict. In the twentieth century, ecological disruption became a problem of global scale, but nationalism and national sovereignty have remained entrenched, with nation states inclined to represent above all their economic interests on the international stage (Brown and Garver 2009; Capra and Mattei 2015). "Ecological degradation is influenced by the structure and dynamics of the world capitalist system, arising from the fact that a single world economy is divided into numerous nation-states, competing with each other both directly and via their corporations" (Clark and Foster 2009: 312). Moreover, States' interest in promoting ecologically harmful economic activity that provides royalties or tax revenues needed for government programs is fundamentally in conflict with their interest in protecting the environment (M'Gonigle and Takeda 2013).

National and international legal systems reflect the strong global emphasis on State sovereignty (Capra and Mattei 2015). National labor and environmental standards have been politically negotiated within capitalist and other contexts, but always with primacy for economic growth, only rarely with critical ecological limits in mind, and never to control aggregate ecological impacts sufficiently (McNeill 2000; Speth 2008; Garver 2013; M'Gonigle and Takeda 2013). Growth-insistent economics in the context of international competition lead to laws that undervalue future generations and the intrinsic value of non-human life and life systems (Costanza and Daly 1987; Martinez-Alier 1987; Daly 1996). At the international stage, the entrenchment of national sovereignty has prevented widespread enactment of effective global controls on climate change, biodiversity loss and other global ecological problems, with the exception of the Montreal Protocol, which looks on track to restore the stratospheric ozone layer (Victor 2008; Brown and Garver 2009; Pelletier 2010; Garver 2013; Capra and Mattei 2015).

Governments know how to adopt rigorous enforcement regimes that transcend State sovereignty if they want to. For example, Canada, Mexico and the US did so in Chapter 11 of the *North American Free Trade Agreement (NAFTA)*[2] by waiving their sovereign immunity so as to allow private investors to seek monetary awards for breach of the *NAFTA*'s investor protections through binding arbitration. Dispute settlement provisions in the *NAFTA* and its environmental side agreement, the *North American Agreement on Environmental Cooperation (NAAEC)*, for resolving concerns regarding weak environmental protection or enforcement are at best weak cousins to Chapter 11's investor dispute mechanism (Tollefson 2002). How law regarding trade might ensure a more enduring, mutually enhancing human-Earth relationship is the subject of Chapter 13 in Part IV of this book.

Europe is an important exception to State sovereignty that impedes an effective supranational, or global, legal order for the human-Earth system. It is both the birthplace of the State-centered Westphalian model and, with the European Union, the most significant contemporary example of a supranational structure that transcends State sovereignty systemically. The supranational supremacy of European law in key areas emerged as an essential characteristic of the 1957 Treaty of Rome and its progeny, with the European Court of Justice playing a determinative role in establishing areas where European law is supreme.[3] Although protection of the environment was not a core objective when the Treaty of Rome established the European Community, it emerged, as the European Court of Justice recognized, as implicit in the Community's authority to regulate the functioning of Europe's common market (Koppen 2005).

The European example may provide a model for a similar evolution of judicially enforceable supranational authority to environmental matters, guided by the principles of subsidiarity and proportionality that play a prominent role under European Union treaties. Already, a global regime of enforceable supranational rules exists in the arena of trade. Article 5(3) of the *Treaty on European Union* [TEU] states:

> Under the principle of subsidiarity, in areas which do not fall within its exclusive competence, the Union shall act only if and in so far as the objectives of the proposed action cannot be sufficiently achieved by the Member States, either at central level or at regional and local level, but can rather, by reason of the scale or effects of the proposed action, be better achieved at Union level.

Thus, subsidiarity is a principle that favors intervention at the level at which it will be most effective for achieving policy objectives (Saunier and Meganck 2009). In practice, the question of who decides what that level is—the central authority, the lower level authority or perhaps both—can lead to contention as to how the principle is applied. Proportionality, reflected in Article 5(4) of the TEU, complements subsidiarity with the notion that governments at all levels must have sufficient capacity and authority to achieve their mandates and objectives (Brown and Garver 2009).

Property rights and other individual freedoms

Modern notions of property, in the sense of a socially recognized and legally protected entitlement to possess or use a good, idea, service or other holder of value, can be traced back to the social stratification, economic specialization and power inequalities that emerged in permanent settlements that arose in the Neolithic Transition (Ponting 2007). Excludability and rivalness are core features of different kinds of property (Daly and Farley 2004). Property can be universally shared, as with air, which everyone can breathe without excluding others; shared exclusively within a defined group, as with common pool resources (CPR) (Ostrom 1990) or controlled exclusively by an individual or other legally recognized entity that can prevent others from using it. As well, some things, such as a car, are rival in that possession or use by one individual excludes possession or use by others. Other things, such as many public goods like clean air, are not. Whereas rivalness derives from inherent characteristics of an item of property, excludability is generally a social construct governed by some kind of legal system.

The way in which the legal order governs the rights of exclusion related to various kinds of property (private, public, common) is a key factor in the human-Earth relationship (Grinlinton and Taylor 2011). On one hand, starting with Garret Hardin's proposal for addressing the Tragedy of the Commons, property rights may help limit overexploitation of some kinds of open access resources (Hardin 1968; Rose 1998).[4] The idea is that

> [w]ith property, each individual harvests the rewards of her care and effort in the management of her resources, just as she suffers the losses from her sloth and poor management; those features of property make her vastly more likely to exercise diligence and prudence about the things she owns.
>
> (Rose 1998: 130–131)

By this view, the more people privatize and commodify, the better stewards they will be—and the richer, too. This is the prevailing view. "From the demise of the authoritarian socialist regimes, we have taken the lesson that modern economies need not the centralization of direct governmental control, but rather the decentralization associated with property and contract" (Rose 1998: 130).

On the other hand, a property-based approach does not easily apply to public goods like clean air, a safe climate, abundant biodiversity and other environmental values, and in fact can make them worse (Hardin 1968). The tendency of the property owner to have "difficulty seeing why it is not his natural right to muddy the waters flowing past his door" (Rose 1998: 130) is at the heart of the growing global ecological crisis. The increasing trend toward monetization of ecosystem services and natural capital (Costanza et al. 1997) exacerbates this problem by expanding the arenas for transactions in which property rights can be exchanged beyond traditional market settings and local contexts (Daly and Farley 2004). The incremental "muddying" of the global economy's innumerable "waters" by each of the myriad consumers and producers, big and small, individual or corporate, who make up the human enterprise adds up to an aggregate level of ecological degradation—a surpassing of global boundaries—that the legal system has so far not managed to address adequately anywhere on Earth.

Prevailing international and national regimes of property rights are an essential feature of the globalized economic system. Strong notions of private property rights that originated with John Locke in the seventeenth century and that were reinforced in the French Revolution have persisted to the present (Ost 2003; Capra and Mattei 2015). Locke's argument for strong protection of private property rights is rooted in the Judeo-Christian notion that humans are superior to and separate from all of God's other creations, that God gave Earth to humans and that private property allows humans to make the most of this gift from God by investing their labor in the land they own and control and reaping the benefits (Capra and Mattei 2015).

In the United States, longstanding principles of common law, along with constitutional and positive law, give private property rights strong protection, but with some restrictions. In the American system and others, government restrictions on private property rights, including environmental regulations, can trigger entitlements to compensation if they deprive property of economic value to a sufficient degree. However, there are limits. For example, in common law, uses of property that create nuisances or trespasses may be enjoined or subjected to claims for damages. The United States Supreme Court has said that although property rights may be compensable if "taken" in various ways, compensation is not required for limitations on uses of property that are imposed in accordance with longstanding principles of common law.[5]

Regulatory measures that harm the value of foreign investments can also trigger investors' claims for compensation under the *NAFTA* and subsequent United States trade agreements, as well as other bilateral or multilateral trade and investment agreements around the world. These rules open avenues for forcing signatory States to pay compensation from public funds, which can chill the willingness to regulate (Byrne 1995; Kmiec 1995; Staff and Lewis 2003). I analyze these investor dispute settlement mechanisms and other aspects of international trade and finance law built on strong notions of private property rights further in Part IV of this book.

Conclusion

Law and other normative social constructs reinforce a dominant global orientation of human society toward sustained economic growth. The conceptual framing of this orientation gives priority to strong private property rights, monetization of non-market goods and services, State sovereignty and a resistance to regulation that is perceived to have potential to interfere too heavily in markets or to limit profits and growth. In parallel with the rise and global spread of this conceptual paradigm from its roots in the European Enlightenment, humans have steadily increased the throughput of material and energy in human society. This rise in consumption has led to increasing aggregate global ecological impacts, to the point where the capacity of Earth systems to support the human enterprise in the long term is more and more in question. In Part II, I explain in more detail why contemporary environmental law is failing to adequately regulate this growing crisis.

Notes

1 *National Environmental Policy Act*, 42 U.S.C. §§ 4321 *et seq*. (2006) (emphasis added).
2 *North American Free Trade Agreement Between the Government of Canada, the Government of Mexico and the Government of the United States*, 17 December 1992, Can T.S. 1994 No. 2, 32 I.L.M. 289 (entered into force 1 January 1994) [*NAFTA*].
3 In *Costa v. ENEL*, C-6/64, [1964] E.C.R. 585 at 593, the Court said:

> By contrast with ordinary international treaties, the EEC Treaty has created its own legal system which, on the entry into force of the Treaty, became an integral part of the legal systems of the Member States and which their courts are bound to apply. By creating a Community of unlimited duration, having its own institutions, its own personality, its own legal capacity and capacity of representation on the international plane and, more particularly, real powers stemming from a limitation of sovereignty or a transfer of powers from the States to the Community, the Member States have limited their sovereign rights, albeit within limited fields, and have thus created a body of law which binds both their nationals and themselves.

4 Hardin's take on the commons has been criticized. For example, Berkes (2009: 262) noted:

> Hardin's own example of the imaginary English pasture was historically incorrect. The medieval English commons were generally used under locally-devised regulations; for example, "stinting" rules limited the number of heads of animals that each owner was allowed to graze on the village pasture. Medieval English commons operated successfully for many centuries, and several economic historians and other scholars have questioned if a "tragedy" of the sort described by Hardin ever occurred widely [reference omitted].

5 *Palazzolo v. Rhode Island*, 533 U.S. 606, 629 (2001).

References

Beddoe, R., R. Costanza, J. Farley, E. Garza, J. Kent, I. Kubiszewski, L. Martinez, T. McCowen, K. Murphy, N. Myers, Z. Ogden, K. Stapleton and J. Woodward. 2009. Overcoming systemic roadblocks to sustainability: the evolutionary redesign of worldviews, institutions, and technologies. *Proceedings of the National Academy of Science* 106(8): 2483–2489.

Berkes, F. 2009. Revising the commons paradigm. *Journal of Natural Resources Policy Research* 1(3): 261–264.

Berry, T. 1999. *The Great Work: Our Way Into the Future*. New York: Three Rivers Press.

Bosselmann, K. 2008. *The Principle of Sustainability: Transforming Law and Governance.* Burlington VT: Ashgate Publishing Co.

Boulding, K. 1975. Love and Lifeboats. Chapel Talk at Gstavs Adolphs College, St. Peter, Minnesota (November 18, 1975).

Boyd, D.R. 2004. Sustainability Law: (R)Evolutionary directions for the future of environmental law. *Journal of Environmental Law and Practice* 14: 357–384.

Brooks, R.O., R. Jones and R.A. Virginia. 2002. *Law and Ecology: The Rise of the Ecosystem Regime.* Burlington VT: Ashgate Publishing Co.

Brown, P.G., and G. Garver. 2009. *Right Relationship: Building a Whole Earth Economy.* San Francisco: Berrett-Koehler.

Bugge, H.C. 2013. Twelve Fundamental Challenges in Environmental Law, Chapter 1 in C. Voigt, ed., *Rule of Law for Nature: New Dimensions and Ideas in Environmental Law*. New York: Cambridge University Press.

Burdon, P.D. 2015. *Earth Jurisprudence: Private Property and the Environment.* New York: Routledge.

Burke, E. III and K. Pomeranz, eds. 2009. *The Environment and World History.* Los Angeles: University of California Press.

Byrne, J.B. 1995. Ten arguments for the abolition of the regulatory takings doctrine. *Ecology Law Quarterly* 22: 89–142.

Capra, F., and U. Mattei. 2015. *The Ecology of Law: Toward a Legal System in Tune with Nature and Community.* San Francisco: Berrett-Koehler.

Clark, B., and J.B. Foster. 2009. Ecological imperialism and the global metabolic rift: unequal exchange and the guano/nitrates trade. *International Journal of Comparative Sociology* 50(30–34): 311–334.

Costanza, R., and H.E. Daly. 1987. Toward an ecological economics. *Ecological Modeling* 38: 1–7.

Costanza, R., R. d'Arge, R. de Groot, S. Farber, M. Grasso, B. Hannon, K. Limburg, S. Naeem, R.V. O'Neill, J. Paruelo, R.G. Raskin, P. Sutton and M van den Belt. 1997.

The value of the world's ecosystem services and natural capital. *Nature* 387: 253–260.

Crosby, A.W. 1987. The Columbian Voyages, the Columbian Exchange, and Their Historians. *Essays on Global and Comparative History*. Washington: American Historical Association.

Cullinan, C. 2011. *Wild Law: A Manifesto for Earth Justice (2nd ed.)*. White River Junction VT: Chelsea Green.

Daly, H.E. 1996. *Beyond Growth*. Boston: Beacon Press.

Daly, H.E., and J. Farley. 2004. *Ecological Economics: Principles and Applications*. Washington, DC: Island Press.

Davies, G. 2002. *A History of Money from Ancient Times to the Present Day*. Cardiff: University of Wales Press.

Delgado, R. 1991. Our better natures: a revisionist view of Joseph Sax's public trust theory of environmental protection, and some dark thoughts on the possibility of law reform. *Vanderbilt Law Review* 44(6): 1209–1227.

Diamond, J. 2005. *Collapse: How Societies Choose to Fail or Succeed*. New York: Viking Press.

Dow, G.K., C.G. Reed and N. Olewiler. 2009. Climate reversals and the transition to agriculture. *Journal of Economic Growth* 14: 27–53.

Earth Charter. 2020. Earth Charter website [online]: http://earthcharter.org.

Elliott, H. 2005. *Ethics for a Finite World: An Essay Concerning a Sustainable Future*. Golden CO: Fulcrum Publishing.

Escobar, A. 2010. Latin America at a crossroads. *Cultural Studies* 24(1): 1–65.

G20. 2010. *The G20 Seoul Summit Leaders' Declaration* [online]: www.g20.org/Documents2010/11/seoulsummit_declaration.pdf

G20. 2011. *Cannes Summit Final Declaration* [online]: www.g20.org/Documents2011/11/Cannes%20Declaration%204%20November%202011.pdf

G20. 2013. *G20 Leaders' Declaration, Saint Petersburg Summit* [online]: www.g20.org/sites/default/files/g20_resources/library/Saint_Petersburg_Declaration_ENG_0.pdf

G20. 2014. *G20 Leaders' Communiqué, Brisbane Summit* [online]: https://g20.org/wp-content/uploads/2014/12/brisbane_g20_leaders_summit_communique1.pdf

G20. 2015. *G20 Leaders' Communiqué, Antalya Summit* [online]: www.g20.utoronto.ca/2015/151116-communique.html

G20. 2019. *G20 Osaka Leaders' Declaration* [online]: www.g20.utoronto.ca/2019/2019-g20-osaka-leaders-declaration.html

Garver, G. 2013. The rule of ecological law: the legal complement to degrowth economics. *Sustainability* 5(1): 316–337 [online]: www.mdpi.com/2071-1050/5/1/316/htm

Garver, G. 2015. Forgotten Promises: Neglected Environmental Provisions of the NAFTA and the North American Agreement on Environmental Cooperation, book chapter in H. Kong and K. Wroth, eds., *NAFTA and Sustainable Development*. New York: Cambridge University Press.

Garver, G. 2019. Confronting remote ownership problems with ecological law. *Vermont Law Review* 43(3): 425–454.

Georgescu-Roegen, N. 1971. *The Entropy Law and Economic Process*. Cambridge MA: Harvard University Press.

Gowdy, J.M., and C.N. McDaniel. 1999. The physical destruction of Naura: an example of weak sustainability. *Land Economics* 75(2): 333–338.

Grinlinton, D., and P. Taylor, eds. 2011. *Property Rights and Sustainability: The Evolution of Property Rights to meet Ecological Challenges*. Boston: Brill.

Hardin, G. 1968. The tragedy of the commons. *Science* 162: 1243–1248.

Heinzerling, L. 1995. Selling pollution, forcing democracy. *Stanford Environmental Law Journal* 14: 300–344.

Hoffman, R.C. 2001. Frontier foods for late medieval consumers: culture, economy, ecology. *Environment and History* 7: 131–167.

Jackson, T. 2009. *Prosperity Without Growth: Economics for a Finite Planet.* Washington, DC: Earthscan.

Kallis, G. 2018. *Degrowth.* Newcastle upon Tyne, UK: Agenda Publishing.

Klein, N. 2014. *This Changes Everything: Capitalism v. the Climate.* New York: Knopf Canada.

Kmiec, D.W. 1995. At last, the Supreme Court solves the takings puzzle. *Harvard Journal of Law & Public Policy* 19: 147–159.

Koppen, I.J. 2005. The Role of the European Court of Justice, in A. Jordan, ed., *Environmental Policy in the European Union, Second Edition.* Sterling, VA: Earthscan.

Kornfeld, I.E. 2009. Mesopotamia: A History of Water and Law, Chapter 3 in J. Dellapenna and J. Gupta, eds., *The Evolution of Law and Politics of Water.* New York: Springer.

Korten, D. 2006. *The Great Turning: From Empire to Earth Community.* San Francisco: Berrett-Koehler.

Krausmann, F., S. Gingrich, N. Eisenmenger, K.H. Erb, H. Haberl and M. Fischer-Kowalski, 2009. Growth in global materials use, GDP and population during the 20th century. *Ecological Economics* 68: 2696–2705.

Kysar, D.A. 2002. Law, environment, and vision. *Northwestern University Law Review* 97: 675–729.

Leopold, A. 1949. *A Sand County Almanac.* New York: Oxford University Press.

Martinez-Alier, J. 1987. *Ecological Economics: Energy, Environment and Society.* Cambridge MA: Basil Blackwell Inc.

McLellan, R., L. Iyengar, B. Jeffries and N. Oerlemans, eds. 2014. *Living Planet Report 2014: Species and Spaces, People and Places.* Gland, Switzerland: WWF.

McNeill, J.R. 2000. *Something New Under the Sun: An Environmental History of the Twentieth-Century World.* New York: W.W. Norton & Co.

McNeill, J.R. 2003. Observations on the nature and culture of environmental history. *History and Theory* 42(4): 5–43.

Merchant, C. 1980. *The Death of Nature: Women, Ecology and the Scientific Revolution.* San Francisco: Harper & Row.

Moore, J.W. 2003. Capitalism as world-ecology: Braudel and Marx on environmental history. *Organization & Environment* 16(4): 431–458.

Moore, J.W. 2007. Silver, Ecology, and the Origins of Modern World, 1450–1640, pp. 123–142 in A. Hornborg, J.R. McNeill and J. Martinez-Alier, eds., *Rethinking Environmental History: World-system History and Global Environmental Change.* Toronto: AltaMira Press.

M'Gonigle, M. 2008. Green legal theory: a new approach to the concept of environmental law. *Neue Konzepte* 4: 34–38.

M'Gonigle, M., and L. Takeda. 2013. The liberal limits of environmental law: a green legal critique. *Pace Environmental Law Review* 30(3): 1005–1115.

Nadeau, R.L. 2006. *The Environmental Endgame: Mainstream Economics, Ecological Disaster, and Human Survival.* New Brunswick NJ: Rutgers University Press.

Naess, A. 1984. A defence of the deep ecology movement. *Environmental Ethics* 6: 265–270.

Nyquist, B., A. Persson, F. Moberg, L. Persson, S. Cornell and J. Rockström. 2013. *National Environmental Performance on Planetary Boundaries: A Study for the Swedish Environmental Protection Agency.* Stockholm: Swedish Environmental Protection Agency.

Oreskes, N., and E.M. Conway. 2010. *Merchants of Doubt.* New York: Bloomsbury Press.

Ost, F. 2003. *La nature hors la loi: L'écologie à l'épreuve du droit.* Paris: La Découverte/Poche.

Ostrom, E. 1990. *Governing the Commons: The Evolution of Institutions for Collective Action.* New York: Cambridge University Press.

Pearce, F. 2012. *The Land Grabbers: The New Fight Over Who Owns the Earth.* Boston: Beacon Press.

Pelletier, N. 2010. Of laws and limits: an ecological economic perspective on redressing the failure of contemporary global environmental governance. *Global Environmental Change* 20: 220–228.

Polanyi, K. 1957. *The Great Transformation: The Political and Economic Origins of Our Time*. Boston: Beacon Press.

Pomeranz, K. 2002. Political economy and ecology on the eve of industrialization: Europe, China, and the global conjuncture. *The American Historical Review* 17(2): 425–446.

Pomeranz, K. 2009. Introduction: World History and Environmental History, pp. 3–32 in E. Burke III and K. Pomeranz, eds., *The Environment and World History.* Los Angeles: University of California Press.

Ponting, C. 2007. *A New Green History of the World: The Environment and the Collapse of Great Civilizations.* Toronto: Penguin Books.

Prince, J.D. 1904. The code of Hammurabi. *The American Journal of Theology* 8(3): 601–609.

Rees, W.E. 2002. Globalization and sustainability: conflict or convergence? *Bulletin of Science, Technology & Society* 22.

Rengasamy, P. 2006. World salinization with emphasis on Australia. *Journal of Experimental Botany* 57(5): 1017–1023.

Richards, J.F. 2003. *The Unending Frontier: An Environmental History of the Early Modern World.* Berkeley CA: University of California Press.

Richards, J.F. 2009. Toward a Global System of Property Rights in Land, pp. 54–78 in E. Burke III and K. Pomeranz, eds., *The Environment and World History.* Los Angeles: University of California Press.

Rose, C.M. 1998. The several future of property: of cyberspace and folk tales, emission trades and ecosystems. *Minnesota Law Review* 83: 129–182.

Ross, A. 1966. A Critique of the Philosophy of Natural Law, pp. 68–74 in M.R. Golding, ed., *The Nature of Law.* New York: Random House.

Ruhl, J.B. 1996. Complexity theory as a paradigm for the dynamical law-and-society system: a wake-up call for legal reductionism and the modern administrative state. *Duke Law Journal* 45(8): 849–928.

Saunier, R.E., and R.A. Meganck. 2009. *Dictionary and Introduction to Global Environmental Governance, Second Edition.* London: Earthscan.

Scott, J.C. 1998. *Seeing Like a State.* New Haven CT: Yale University Press.

Sen, A. 1999. *Development as Freedom.* New York: Anchor Books.

Sieferle, R.P. 2001. *The Subterranean Forest: Energy Systems and the Industrial Revolution.* Cambridge UK: The White Horse Press.

Sörlin, S., and P. Warde. 2007. The problem of the problem of environmental history: a re-reading of the field. *Environmental History* 12: 107–130.

Speth, J.G. 2008. *The Bridge at the Edge of the World.* New Haven, CT: Yale University Press.

Staff, M.J., and C.W. Lewis. 2003. Arbitration under NAFTA Chapter 11: past, present, and future. *Houston Journal of International Law* 25: 301–340.

Tarlock, A.D. 2000. The future of environmental "rule of law" litigation. *Pace Environmental Law Review* 17: 237–272.

Testart, A. 1982. The significance of food storage among hunter-gatherers: residence patterns, population densities, and social inequalities. *Current Anthropology* 23(5): 523–537.

Tollefson, C. 2002. Games without frontiers: investor claims and citizen submissions under the NAFTA regime. *Yale International Law Journal* 27: 141–191.

United Nations. 1987. *Report of the World Commission on Environment and Development: Our Common Future.* New York: United Nations.

United Nations. 2015a. *Transforming Our World: The 2030 Agenda for Sustainable Development.* New York: United Nations.

United Nations Environment Programme (UNEP). 2011. *Towards a Green Economy: Pathways to Sustainable Development and Poverty Eradication—A Synthesis for Policymakers.* Nairobi: UNEP. www.unep.org/greeneconomy/GreenEconomyReport/tabid/29846/language/en-US/Default.aspx

Victor, P. 2008. *Managing Without Growth.* Cheltenham, UK: Edward Elgar.

von Weizsäcker, E., K. Hargroves, M.H. Smith, C. Desha and P. Stasinopoulos. 2009. *Factor Five: Transforming the Global Economy through 80% Improvements in Resource Productivity.* London: Earthscan.

Westra, L. 2011. Governance for Integrity? A Necessary but Distant Goal, pp. 324–336 in P. Burdon, ed., *Exploring Wild Law: The Philosophy of Earth Jurisprudence.* Kent Town, S. Australia: Wakefield Press.

Part II

The failure of environmental law to respect key ecological limits

5 An overview of the failures of environmental law

Contemporary environmental law is fully embedded in the legal system that evolved along the long historical path to the Anthropocene reviewed in Part I. In operation, environmental law and ecological law must both exist within an institutional structure consisting of "a set of principles, rules, norms and procedures" along with physical and organizational infrastructure (Ivanova 2010: 37). The existing global complex of environmental institutions, from the global to the local level, has hardly prevented global ecological threats from worsening, and they are unlikely to do so without radical reform (Ivanova 2010: 30; Pelletier 2010).

Ecological law is emerging as a response to the failure of contemporary environmental law, and the broader legal system, to address the ecological dilemma that the Anthropocene represents. The 2016 *Oslo Manifesto for Ecological Law and Governance*[1] notes that despite the objective to adequately protect the environment, "environmental law has failed to halt ecological degradation and continues to fall short of its objectives." According to the manifesto, ecological law aims to overcome key interrelated flaws of contemporary environmental law: first, its grounding in outdated anthropocentric, Cartesian worldviews that promote individualism, utilitarianism and human dominance over nature; and second, its reductionist, as opposed to holistic, tendency to focus on environmental problems as discrete and isolated.

Anthropocentrism and the tendency toward monetization in environmental law

A core manifestation of anthropocentrism, individualism and utilitarianism in environmental law is the tendency toward monetization as a way to normalize social preferences and to regulate the relationship between environmental problems and the development pressures that create them, for example by assessing the costs and benefits of environmental regulation primarily in terms of monetized valuations (Heinzerling 2005: 164–169). The tendency to incorporate the language of money into the law is rooted ultimately in the conception of humans as apart from nature, and as rational actors free to accommodate and own elements of nature in the quest to maximize personal wealth and well-being (Nadeau 2006: 81–123).

The assumption that all relevant factors in the economy can be monetized and then subject to economic forces has even crept into ecological economics, along with other fundamentals of a neo-classical economic perspective. In 1997, a prominent ecological economist led a group of researchers who estimated the total value of the world's ecosystem services and natural capital at $33 trillion, compared to a world gross national product of $18 trillion (Costanza et al. 1997). Building on this concept, the *Synthesis Report of The Economy of Ecosystems and Biodiversity* (*TEEB Report*) estimated the value of pollination by bees and other insects globally at 153 billion euro (TEEB 2010). This figure is perhaps intended to express in the most ideal, universal form, the importance of bees. However, it dangerously implies that if bees disappeared as a result of human activities, for 153 billion euro we could purchase a substitute for them.

Yale economist William Nordhaus does something similar with his Dynamic Integrated model of Climate and the Economy [DICE], which he uses to "weigh[] the costs of slowing climate change against the damages of more rapid climate change" (Nordhaus 2007: 10), with all the costs and benefits expressed in dollars. The model combines models of the economy based on neo-classical economics with geophysical models of the global carbon cycle, climate change and damages to society from climate change (Nordhaus 2007: 27–33). He concludes that with annual average global economic growth of approximately two percent between 2000 and 2100, and discounting future value at four percent annually, global society should invest $2 trillion over the next century to eliminate $5 trillion of climate damages, leaving $17 trillion of climate damages unabated because we would be better off investing in other things that will have a higher net benefit (Nordhaus 2007). Nordhaus assumes[2] that the value of the vast ecosystem impacts of climate change can be monetized and compared readily to the values of goods and services in the economy, and that society will be able to buy its way out of whatever problems are related to an average global temperature 2.6 °C higher than in pre-industrial times—the global heating in his optimal scenario (Nordhaus 2007: 167). Nordhaus has updated the model regularly since first creating it in 1992, with relatively minor and major changes to some elements, but its basic structure—and its expression of climate damages in monetary terms—remains the same (Nordhaus 2018).

Monetary valuation of ecosystems and their services to humans, or of anything really, carries with it the assumption that having assigned a dollar value, we have discerned a thing's true value. Yet, in doing so, we subject that thing to the impossible assumptions of perfect information and idealized rationality in the market and to the increasing abstractness of money and wealth, and we strip it of its ecological—and, ultimately, spiritual—essence as a process or an element that triggers and responds to reactions in a complex, evolving web of ecological relationships. We treat the monetized thing as an object, not as a fellow member of Earth's life community (Berry 1999).

Tortured comparisons of value as in the *TEEB Report* and Nordhaus's work using the DICE model support a legal infrastructure that includes contemporary environmental law. Under ecological law, the value of pollination by bees would be assessed on the basis of such factors as bees' needs for their share of the Earth's ecological capacity, along with their role in the functioning of ecosystems, including in some cases their functioning so as to provide food to people and other living beings. The legal infrastructure would support a focus on understanding the role of bees in maintaining ecosystem functioning and on limiting the impacts of the human economy on bees and their role in ecosystem functioning to the minimum level that allows both bees and humans to flourish. But human flourishing in this conception does not imply unlimited consumption and affluence; rather, it is flourishing within a space for humans collectively that is bounded by ecological limits and that allows flourishing of as much other life as possible. Attaching a dollar value to pollination by bees works counter to this framework—and, it is insulting to bees.

Ecological law will undoubtedly impose on human activities limitations that do not exist under the current legal regimes in much of the developed world. A system in which ecological restraints envelop the economy requires the collectivity of economic actors to limit their choices so that, taken together, they respect those ecological limits. Yet, in the market economies of both wealthy and less wealthy nations, the freedom to spend on activities and maximize personal wealth, often regardless of their ecological costs, is paramount (Rees 2002: 251). The reward that a well-compensated person expects is not just money; it is what that money can purchase: often, one or more large homes, jet-fueled vacations in far-off places and a host of other material and energy-intensive luxuries. This legally protected consumption, rooted in strong notions of property rights and personal freedom, exacts ecological costs that are collectively shared and often incremental, diffuse and delayed and consequently external to the economic and legal infrastructure. Analyses of the trade flows of material and energy resources indicate that their consumption in high-income countries contributes significantly to biodiversity losses that are most severe in low- and middle-income countries (Foley et al. 2007; Pollard et al. 2010). The current institutional infrastructure for law, including environmental law, does little to address these ecological stresses.

In short, ecological law would cordon off the sphere of market exchanges based on abstract monetary units to a considerably greater extent than occurs under contemporary notions of environmental law. Ecological law flows from the notion of enclosing the spark of economic competition and market exchange within an engine of cooperation that runs according to the dynamics of the global ecosystem (Brown and Garver 2009: 86–87). More deeply, it draws from "a feeling of awe for the cosmos and embracing an ethic of humankind's appropriate place in, and relationship to, the cosmos and the earth" (Brown and Garver 2009: 20).

Reductionism and lack of holistic perspectives in environmental law

The reductionist approach in environmental law obscures consideration of the aggregate impacts of human activity on the global ecosphere (Boyd 2004: 366) and impedes the full integration of a systemic ecological approach into the entire legal infrastructure. Although, as noted earlier in Part I, the holistic approach has not been entirely absent from contemporary environmental law, particularly in the 1960s and 1970s, which gave way to the economically-driven ideology of the 1980s, and the economic gloss over environmental law has only hardened since.

The reductionist approach is evident in "global" environmental law that transcends traditional understandings of international, domestic and local law. Yang and Percival (2009: 616–617) describe global environmental law as follows:

> Global environmental law is the set of legal principles developed by national, international, and transnational environmental regulatory systems to protect the environment and manage natural resources. As a body of law, it is made up of a distinct set of substantive principles and procedural methods that are specifically important or unique to governance of the environment across the world. It includes: (1) public international environmental law, commonly used to refer to the set of treaties and customary international legal principles governing the relations between nations; (2) national environmental law, which describes the principles used by national governments to regulate the behavior of private individuals, organizations, and subnational governmental entities within their borders; and (3) transnational law, which describes the set of legal principles used to regulate the cross-border relationships between private individuals and organizations.

Prominent in this notion of global environmental law are the precautionary principle and the "polluter pays" principle, along with widespread use of environmental impact assessments and permit systems for polluters, all of which support the central substantive goals of protecting human health and the integrity of ecosystems (Yang and Percival 2009: 623).

Yang and Percival (2009: 623–626) provide a comprehensive description of the global complex of soft and hard environmental legal regimes, which are implemented through both public and private initiatives. This emerging global environmental law reflects a stark schism in the law with respect to the most pressing global ecological challenges: an increasingly interlinked network of approaches for dealing with certain disaggregated environmental stress on one hand, and a persistent failure to deal with the most momentous, aggregate environmental problems on the other. Yang and Percival's analysis invites an assessment of which features of global environmental law are leading to and which are impeding the development of a global legal architecture that supports effective legal limits on human use of the Earth's ecocapacity. Overall, global environmental law as these authors describe it falls short.

The principal problem is that global environmental law is essentially the handmaiden of growth-driven economic globalization. Yang and Percival (2009: 616) present global environmental law as a counterpart to sustained economic growth, which they appear to endorse without contemplating whether such growth is tenable in light of the growing global ecological crisis. Although they describe well the emerging ways in which environmental legal regimes, norms, procedures and policy are transcending traditional notions of State-led international law, they neglect to situate global environmental law within the broader legal and institutional context, with its own emerging trends, that envelops the increasingly globalized economy. In noting "a growing convergence around a few principal approaches to environmental regulation" (Yand and Percival 2009: 616), they fail to ask whether those approaches are capable of reining in the enormous drive to unleash financial capital "to mobilize global resources as fodder for industrial metabolism" (Pelletier 2010: 220), with adequate attention to how the aggregate scale of the economy measures up against global ecological limits.

To remove obstacles and find openings for the integration of ecological law, a comprehensive review is necessary of this broader legal infrastructure from the global to the local level. A prime candidate for such review is the global trade regime. Yang and Percival (2009: 643) contend that trade liberalization, as reflected for example by the *NAFTA* and its side agreements, has promoted legislative reforms that protect the environment. They note that following the adoption of the environmental side agreement to *NAFTA*, which purported to address a range of trade-related environmental concerns,

> post-NAFTA trade liberalization negotiations with other nations, ranging from Jordan to Peru, have led the United States to press for adoption of environmental provisions within those agreements. In other words, trade liberalization can also be used as a tool to spur greater environmental protection.
>
> (Yang and Percival 2009: 643)

Yet, the environmental provisions contained in these trade agreements or their environmental side agreements can also be seen as lost opportunities, if they are not assessed and critiqued based on the post-adoption record and are nonetheless duplicated in one trade agreement after the other. Once in place, they could relieve the pressure to establish stronger, more effective rules. This is in fact what has occurred.

The environmental provisions of many recent trade agreements in the Americas are part of an overall global approach that promotes trade to generate growth (Rees 2002: 251; Pelletier 2010: 224–225), with relatively hard legal provisions to protect investments and punish protectionism but only soft mechanisms aimed at protecting the environment. Although modern trade agreements typically give precedence to at least some multilateral environmental agreements in the case of conflicting provisions or interpretations,[3] in general multilateral environmental agreements are relatively weak, unenforceable and grounded in a hierarchy of values that allow political and economic concerns to trump environmental protections (Pelletier 2010: 221–222). With ecological law, the opposite would be true. Regulation of trade would be oriented at the outset toward driving down the total ecological footprint of trade, using the principles that underlie, for example, the green building movement: minimize demand for energy, water and resources; maximize renewable, low-carbon energy; maximize reuse and recycling of materials; and favor local sourcing over long-range transport of materials unless it does not minimize ecological footprint (Commission for Environmental Cooperation 2008: 16–19).

The interlinked international and national regimes of property rights, discussed earlier in Chapter 4, constitute another aspect of reductionism where review against structural criteria for maintaining the economy within ecological bounds is warranted. In many respects, property rights serve as the glue that keeps the globalized economic system together as it is currently formulated. Yet, no system of property rights is absolute, and longstanding principles of common law, along with constitutional and positive law, put restrictions on the use of property, including environmental ones. For example, the use of property to create nuisances or trespasses may be enjoined or subjected to claims for damages, and the United States Supreme Court has said that compensable property rights exclude protection from liability for common law infractions and other limitations that inhere in property title.[4] On the other hand, in the American system and others, government restrictions on property rights, including environmental regulations, can trigger constitutional entitlements to just compensation if they deprive the property of all but token economic value.[5] Regulatory measures that harm the value of foreign investments can also trigger investors' claims for compensation under recent United States trade agreements. Strong protection of private property rights as in the United States Constitution and modern trade and investment agreements open avenues for forcing public officials into settlements from public funds, which can chill the willingness to regulate (Byrne 1995: 90; Kmiec 1995: 151; Staff and Lewis 2003: 319–320). In a legal and policy environment in which maintaining economic activity within ecological boundaries is paramount, property rights will have incorporated an expanded set of responsibilities and regulatory restrictions for which compensation would not be due.

In sum, contemporary environmental law only marginally moderates the prevailing association of property with a paradise of "peaceful, fruitful effort" that drives the "grand bustle of trade and labor [by which] we all grow richer" (Rose 1998: 131). Ecological law would call into question longstanding notions of property rights rooted in the Anglo-American legal tradition and now prevalent worldwide.

Another area of shortcomings of contemporary environmental law in which reductionism ultimately prevails is environmental impact assessment (EIA), which is one of the most far-reaching links between environmental protection and the broader legal and policy infrastructure. Many legal systems and financial institutions have followed the example of the United States in adopting an EIA requirement like the one in the *National Environmental Policy Act* (*NEPA)*. The "look before you leap" and public participation aspects of EIA are important, and overall EIA has undoubtedly led to better environmental outcomes in many cases. However, EIA laws are typically procedural only; a finding of significant environmental impacts virtually never requires rejection of a project or policy.[6] Moreover, governments typically must consider only impacts and future development of a project or policy that are foreseeable and concrete, which limits the extent to which alternatives to, and cumulative and indirect impacts of, the proposed action must be taken into account.[7] Even with strategic EIA requirements, whereby impacts are evaluated at a broad policy or programmatic level, the full picture can fall through the legal cracks.[8] More often, EIA is undertaken in detail at the project level, so that the cumulative impacts are inevitably isolated from the broader perspective. Further, monitoring of impacts following the completion of an EIA is notoriously weak (Kass 2009: 84), at least in the United States, where such monitoring is not always mandatory[9] and is rarely conducted and where the Supreme Court has limited the ability to challenge implementation of a decision based on subsequent monitoring.[10] Supported with adequate research and monitoring, and contained within an infrastructure for ecological law that includes substantive limitations that build off of impact assessments, EIA has an important role to play in maintaining the human economy within global ecological boundaries.

Notes

1 Ecological Law and Governance Association (June 2016) [online]: www.elga.world/oslo-manifesto.
2 Nordhaus's assumptions have received harsh criticism (Sterman 2002: 519).
3 See e.g. *NAFTA* art. 104 (providing that only the *Convention on International Trade in Endangered Species of Wild Fauna and Flora*, done at Washington, March 3, 1973, as amended June 22, 1979; the *Montreal Protocol on Substances that Deplete the Ozone Layer*, done at Montreal, September 16, 1987, as amended June 29, 1990; the *Basel Convention on the Control of Transboundary Movements of Hazardous Wastes and Their Disposal*, done at Basel, March 22, 1989, on its entry into force for Canada, Mexico and the United States; and two border-related bilateral agreements between the US and Canada and the US and Mexico shall prevail over provisions of the *NAFTA* to the extent of inconsistency).
4 *Palazzolo v. Rhode Island*, 533 U.S. 606 at 629 (2001).
5 See *Lucas v. South Carolina Coastal Council*, 505 U.S. 1003 at 1015 (1992); *Agins v. City of Tiburon*, 447 U.S. 255 at 261 (1980).
6 See e.g. *Robertson v. Methow Valley Citizens Council*, 490 U.S. 332 at 350 (1989).
7 See e.g. *Vermont Yankee Nuclear Power Corp. v. Natural Resources Defense Council, Inc.*, 435 U.S. 519 at 551–552 (1978) (limiting the scope of alternatives that agencies are required to consider); *Robertson v. Methow Valley Citizens Council, ibid.* at 355 (noting emphasis on "reasonably foreseeable" impacts); *Center for Biological Diversity v. Nat'l Hwy. Safety Admin.*, 538 F.3d 1172 at 1215–1217 (9th Cir. 2008) (discussing *NEPA* requirement to consider "reasonably foreseeable" cumulative impacts).
8 See *Kleppe v. Sierra Club*, 427 U.S. 390 at 412–415 (1976).
9 See 40 C.F.R. §§ 1505.2(c), 1505.3.
10 See *Norton v. Southern Utah Wilderness Alliance*, 542 U.S. 55 at 72–73 (2004).

References

Berry, T. 1999. *The Great Work: Our Way Into the Future.* New York: Three Rivers Press.

Boyd, D.R. 2004. Sustainability law: (R)Evolutionary directions for the future of environmental law. *Journal of Environmental Law and Practice* 14: 357–384.

Brown, P.G., and G. Garver. 2009. *Right Relationship: Building a Whole Earth Economy.* San Francisco: Berrett-Koehler.

Byrne, J.B. 1995. Ten arguments for the abolition of the regulatory takings doctrine. *Ecology Law Quarterly* 22: 89–142.

Commission for Environmental Cooperation (CEC). 2008. *Green Building in North America: Opportunities and Challenges.* Montreal: Commission for Environmental Cooperation.

Costanza, R., R. d'Arge, R. de Groot, S. Farber, M. Grasso, B. Hannon, K. Limburg, S. Naeem, R.V. O'Neill, J. Paruelo, R.G. Raskin, P. Sutton and M van den Belt. 1997. The value of the world's ecosystem services and natural capital. *Nature* 387: 253–260.

Foley, J.A., C. Monfreda, N. Ramankutty and D. Zaks. 2007. Our share of the planetary pie. *Proceedings of the National Academy of Science* 104(31): 12585–12586.

Heinzerling, L. 2005. Risking it all. *Alabama Law Review* 57: 103–117.

Ivanova, M. 2010. UNEP in global environmental governance: design, leadership, location. *Global Environmental Politics* 10(1): 30–59.

Kass, M.J. 2009. A NEPA climate paradox: taking greenhouse gases into account in threshold significance determinations. *Indiana Law Review* 42: 47–96.

Kmiec, D.W. 1995. At last, the Supreme Court solves the takings puzzle. *Harvard Journal of Law & Public Policy* 19: 147–159.

Nadeau, R.L. 2006. *The Environmental Endgame: Mainstream Economics, Ecological Disaster, and Human Survival.* New Brunswick NJ: Rutgers University Press.

Nordhaus, W. 2007. *The Challenge of Global Warming: Economic Models and Environmental Policy.* New Haven, CT: William Nordhaus.

Nordhaus, W. 2018. Evolution of modeling of the economics of global warming: changes in the DICE model, 1992–2017. *Climatic Change* 148: 623–640.

Pelletier, N. 2010. Of laws and limits: an ecological economic perspective on redressing the failure of contemporary global environmental governance. *Global Environmental Change* 20: 220–228.

Pollard, D., R. Almond, E. Duncan, M. Grooten, L. Hadeed, B. Jeffries and R. McLellan, eds. 2010. *Living Planet Report 2010.* Gland, Switzerland: World Wide Fund for Nature.

Rees, W.E. 2002. Globalization and sustainability: conflict or convergence? *Bulletin of Science, Technology & Society* 22.

Rose, C.M. 1998. The several future of property: of cyberspace and folk tales, emission trades and ecosystems. *Minnesota Law Review* 83: 129–182.

Staff, M.J., and C.W. Lewis. 2003. Arbitration under NAFTA Chapter 11: past, present, and future. *Houston Journal of International Law* 25: 301–340.

Sterman, J.D. 2002. All models are wrong: reflections on becoming a systems scientist. *System Dynamics Review* 18(4): 501–531.

TEEB. 2010. *The Economics of Ecosystems and Biodiversity: Mainstreaming the Economics of Nature: A synthesis of the approach, conclusions and recommendations of TEEB.*

Yang, T., and R.V. Percival. 2009. The emergence of global environmental law. *Ecology Law Quarterly* 36: 615–664.

6 Key examples of the inadequacy of environmental law

Three areas of United States environmental law and one of European environmental law demonstrate the difficulties in implementing strict system-based limits under the contemporary notion of environmental law. All these examples are implicitly or explicitly founded on system-based limits to the introduction of pollutants or on economic activity. The first example is Total Maximum Daily Loads [TMDLs] under the United States' *Clean Water Act*,[1] which establish the maximum amount of a pollutant that can be emitted to a water body without exceeding the federal water quality standards for the pollutant. The second is the as-yet largely undeveloped authority to regulate carbon dioxide through provisions of the United States' *Clean Air Act*[2] so as to mitigate climate change. The third is the critical loads and levels concept employed in the European Union's air pollution regulation[3] and the *Convention on Long-range Transboundary Air Pollution* [*LRTAP*],[4] which is central to air pollution control in Europe. The last is the Acid Rain Program in the United States, specifically the provisions of the *Clean Air Act* that establish a cap-and-trade program for emissions of sulfur dioxide. These mechanisms are described as weak because they have not achieved the systems-based objectives that underlie them. Further, the systems limits on which they are based are not founded entirely on ecosystems-based criteria, and they generally have only indirect links to planetary boundaries and to the threat of global systemic catastrophe. By contrast, the *Montreal Protocol on Substances that Deplete the Ozone Layer* [*Montreal Protocol*][5] is a relatively successful example of a regulation based on broad, global-scale ecological limit.

TMDLs under the United States *Clean Water Act*

The United States *Clean Water Act* requires States to identify waters within their boundaries in which any applicable water quality standard (WQS) required by the Act is not being met and to rank the waters "taking into account the severity of the pollution and the uses to be made of the waters."[6]

> Each State shall establish for the waters identified [as not meeting WQSs], and in accordance with the priority ranking, the total maximum daily load,

> for those pollutants which the Administrator identifies under section 304(a)(2) as suitable for such calculation.[7]

Thus, TMDLs reflect an effort to establish systemic limits tied to ecological and other criteria reflected in water quality standards.

TMDLs must include the pollutant load from point sources that have government-issued permits (the "wasteload allocation"), along with the combined load from nonpoint sources and natural background sources (the "load allocation").[8] They must be set "at levels necessary to attain and maintain the applicable narrative and numerical WQS with seasonal variations and a margin of safety which takes into account any lack of knowledge concerning the relationship between effluent limitations and water quality."[9] Once established, TMDLs provide a basis for incorporating into States' "continuing planning processes"[10] a system for adjusting permits for point sources that contribute to nonattainment of water quality standards and for exploring ways to control nonpoint sources that the *Clean Water Act* regulates much less rigorously (Adler 2009: 10314; Caudill and Curley 2008: note 10).

Unlike end-of-pipe regulations that are established largely on the basis of technological and economic considerations,

> the TMDL program promises an "ambient" approach to water monitoring and standards. That is, instead of focusing on releases from known sources of water pollution (i.e., monitoring discharges from discrete, identifiable pollution sources), regulation and reporting will increasingly be concerned with the *in situ* quality of waterbodies themselves.
>
> (Boyd 2000: 2)

States are required to establish WQS that set limits on ambient water quality, "taking into consideration their use and value for public water supplies, propagation of fish and wildlife, recreational purposes, and agricultural, industrial, and other purposes, and also taking into consideration their use and value for navigation."[11] Thus, WQS under the *Clean Water Act* take into account the uses that States designate for their waters, as well as water quality criteria, which the EPA is required to publish and update, for protecting those designated uses.

Water quality criteria, and hence water quality standards, must reflect the latest scientific knowledge

> (A) on the kind and extent of all identifiable effects on health and welfare including, but not limited to, plankton, fish, shellfish, wildlife, plant life, shorelines, beaches, esthetics, and recreation which may be expected from the presence of pollutants in any body of water, including ground water;
>
> (B) on the concentration and dispersal of pollutants, or their byproducts, through biological, physical, and chemical processes; and

(C) on the effects of pollutants on biological community diversity, productivity, and stability, including information on the factors affecting rates of eutrophication and rates of organic and inorganic sedimentation for varying types of receiving waters.[12]

Water quality standards therefore reflect both the economic considerations that are inherent in designating different uses for different waters and the ecological considerations involved in ensuring that waters are of sufficient quality to protect designated uses.

The EPA's National Water Quality Inventory for the year 2017 indicates that 46 percent of the assessed rivers and streams were "in poor biological condition"; 21 percent of the assessed lakes, ponds and reservoirs were hypereutrophic, mostly because of nitrogen and phosphorous inputs; 18 percent of assessed coastal waters were "in poor biological condition" and 14 percent were "poor based on a water quality index"; and 32 percent of wetlands assessed were "in poor biological condition."[13] As a result, a wide range of waters did not meet WQS, and EPA identified the primary pollution of concern to be mercury, pathogens, nutrients, PCBs, sediment and organic enrichment (causing oxygen depletion) from atmospheric deposition and agricultural run-off—in other words, nonpoint source pollution (USEPA 2017). "Legacy pollution" due to contamination from sediments that release pollutants discharged in the past is another important source of impairment in some waters (Boyd 2000: 5).

Because TMDLs establish the limit on aggregate loadings needed to maintain water quality at levels established, at least in part, according to ecological criteria, they are a possible proxy for a systems-based ecological boundary. They also have at least an indirect relationship to the still largely uncertain planetary boundaries for chemical pollution and biogeochemical flows (nitrogen and phosphorus) (Rockström et al. 2009: 8–9, 12–14, 18–19). However, the States and EPA have been slow in developing TMDLs. The federal and state governments essentially ignored the mandate to establish TMDLs for impaired waters until spurred by litigation to do so in the 1990s (Boyd 2000: 7; Karr and Yoder 2004: 594). Although the States and EPA had established over 70,000 TMDLs as of 2014,[14] even if established, TMDLs are insufficient to overcome other provisions in the *Clean Water Act* that limit or complicate the ability to regulate key contributors to water quality problems, especially nonpoint sources of water pollution.

The functional limitations of the TMDL program are clear. The most critical limitation is that once TMDLs are established, States are not required to develop plans to implement them or to apply them in a binding manner to nonpoint sources. In July 2000, the Clinton Administration EPA finalized a TMDL rule that required States to develop implementation plans providing details on and schedules for actions to reduce point and nonpoint pollutant loadings of impaired waters, but the Bush Administration EPA, responding to criticism that the rule was not based on "sound science," withdrew it in 2003 before it became effective (Caudill and Curley 2008: 256).[15]

Looking forward,

> [f]ederal authority to mandate nonpoint source controls remains weak. Implementation of the analytic tools required by the TMDL process will be costly and difficult. And conflicts are almost certain to arise due to the geographically interrelated nature of pollution sources and legal jurisdictions.
> (Boyd 2000: 4)

Percival (2007: 31) notes that

> [n]on-point source pollution is widely, and properly, understood as one of the top problems that regulatory policy has failed to come to grips with. So long as most land management decisions remain the fiercely guarded province of state and local authorities, solutions to non-point source pollutions problems will remain elusive.

The example of mercury contamination of many United States waters illustrates well the many difficulties in relying on the TMDL program to restore the water quality of impaired waters. In Spring 2020, approximately 4,860 water bodies in the United States were listed as impaired with mercury contamination and the EPA had approved about 6,900 mercury TMDLs.[16] Much mercury impairment of water bodies in the United States is due to atmospheric deposition of mercury, a large share of which comes from the emissions of coal-fired power plants.[17] Identifying which mercury sources affect a particular impaired water body is a complicated matter, although methodologies for establishing these source-receptor relationships are now well advanced. Another complication is that a notable amount of atmospheric deposition of mercury in the United States comes from foreign sources, and a notable amount of mercury emissions in the United States migrate outside of United States borders. Thirty percent of mercury deposition in the Western United States was due to Chinese sources, mostly coal-fired power plants, according to a 2004 estimate, while two-thirds of United States mercury emissions leave the country (Percival 2007: 18–19). As a result of these complications, the United States has far to go before TMDLs can provide a reliable basis for protecting aquatic ecosystems from mercury and other pollutants.

Greenhouse gases under the United States *Clean Air Act*

The United States Supreme Court's ruling in *Massachusetts v. EPA*[18] confirmed the authority of the federal government in the United States to regulate carbon dioxide and other greenhouse gases as pollutants under the *Clean Air Act*. A group of states, local governments and nongovernmental organizations brought the case to challenge the EPA's decision not to regulate greenhouses gases emitted by motor vehicles under the *Clean Air Act* § 202(a)(1). That provision provides:

> The [EPA] Administrator shall by regulation prescribe (and from time to time revise) in accordance with the provisions of this section, standards

> applicable to the emission of any air pollutant from any class or classes of new motor vehicles or new motor vehicle engines, which in his judgment cause, or contribute to, air pollution which may reasonably be anticipated to endanger public health or welfare ....[19]

The Supreme Court's decision turned on its rejection of EPA's argument that substances that contribute to climate change, like greenhouse gases, are not "air pollutants" as defined in the Act. According to the Act,

> [t]he term "air pollutant" means any air pollution agent or combination of such agents, including any physical, chemical, biological, radioactive (including source material, special nuclear material, and byproduct material) substance or matter which is emitted into or otherwise enters the ambient air. Such term includes any precursors to the formation of any air pollutant, to the extent the Administrator has identified such precursor or precursors for the particular purpose for which the term "air pollutant" is used.[20]

The Court held that on its face, this definition unambiguously includes greenhouse gases emitted to the ambient air.[21] As to EPA's determination as to whether an air pollutant endangers the public health of welfare, the Act provides that "effects on welfare includes, but is not limited to, effects on soils, water, crops, vegetation, manmade materials, animals, wildlife, weather, visibility, and climate ... ."[22] The Court ruled that "EPA can avoid taking further action only if it determines that greenhouse gases do not contribute to climate change or if it provides some reasonable explanation as to why it cannot or will not exercise its discretion to determine whether they do."[23]

The decision in *Massachusetts v. EPA* is rooted indirectly in a systems-based analysis of the role of greenhouse gases in altering the global climate system. The Court's reasoning is based in large part on recognition of the potential impacts of pollution not directly on the health of individual humans, but on ecosystems on which human health and welfare depend. Carbon dioxide and the other greenhouse gases stand out in this regard from other pollutants that are regulated under the *Clean Air Act*. Despite this distinction, the Court concluded that carbon dioxide is covered by the *Clean Air Act* even if it is of concern *only* at the global level of climate change, unlike pollutants with more direct and local effects on human health and welfare that are typically regulated under the Act.[24]

Following the decision in *Massachusetts v. EPA*, the EPA in the Obama Administration issued a so-called endangerment finding under *Clean Air Act* §202(a) that emissions of carbon dioxide and five other greenhouse gases may reasonably be anticipated to endanger public health or welfare.[25] The endangerment finding opened the door to broader regulation of greenhouse gases under the *Clean Air Act*. Subsequently, in April 2010, the EPA and the National Highway Traffic Safety Administration [NHTSA] jointly issued new greenhouse gas emission and fuel economy standards for passenger cars and other light-duty motor vehicles in model years 2012 through 2016.[26] In 2012, the EPA and

NHTSA issued a final rule on greenhouse gas emission and fuel economy standards for light-duty motor vehicles in model years 2017 to 2025 that required a fleet average fuel economy of 54.5 miles per gallon by 2025.[27] In 2020, the EPA and NHTSA significantly weakened this requirement in a new rule that only requires light-duty motor vehicles to achieve a fleet average fuel economy of 40.5 miles per gallon by 2030.[28] Neither the endangerment finding or any of these rules makes reference to a global or national limit for greenhouse gas emissions or to safe thresholds for atmospheric concentrations of greenhouse gases, changes in the net radiative forcing or the increase in the average global temperature.

The EPA has undertaken other modest regulatory measures related to climate change. In October 2009, to gather information needed for future policymaking, it finalized a rule mandating as of 2010 the reporting of greenhouse gases from approximately 10,000 power plants and other facilities, which together account for about 85 percent of emissions in the United States. In 2015, the Obama EPA issued a Clean Power Plan rule that was projected to achieve a 32 percent reduction in carbon dioxide emissions from coal-fired power plants by 2030,[29] but the Trump EPA repealed that rule and replaced it in 2019 with a much weaker rule[30] that may increase carbon dioxide emissions from power plants in 2050, compared to having no rule at all (Keyes et al. 2019).

Taken together, EPA's regulatory actions to address climate change include restrictions that in some respects incrementally tend toward, and in significant respects away from, respect for a scientifically derived systemic boundary such as proposed by planetary boundaries researchers. However, EPA's nascent climate change regulations are mute as to what any such boundary ought to be and are ultimately linked to the focus in the *Clean Air Act* on protecting human health, as opposed to ecosystems more broadly. Further, the obstacles to policy mechanisms in the United States that are linked to, and rapidly working toward, a planetary boundary for climate change are significant. Four obstacles stand out.

First, greenhouse emissions targets that the United States has adopted have fallen short of what is needed to meet scientifically derived limits such as those of the planetary boundary for climate change. The United States set a national greenhouse gas emissions target of 26 to 28 percent below 2005 levels under the 2015 *Paris Agreement* of the *United Nations Framework Convention on Climate Change* [UNFCCC],[31] but in November 2019 formally notified the United Nations that it was withdrawing from the agreement.[32] In early 2020, the United States does not have national targets for reducing greenhouse gas emissions. Even the United States' Paris targets did not match up with emissions limits that the IPCC and others deem necessary to meet an eventual return to an atmospheric concentration of carbon dioxide of 350 ppm or a 1.5 or 2 °C global temperature rise[33] above pre-industrial levels (Carbon Action Tracker 2020). The IPCC's 2007 assessment indicates that limiting the global temperature rise to 2 °C will require emissions from developed world countries to decrease by 10 to 40 percent from 1990 levels by 2020 (Gupta et al. 2007: 748). Moreover, even with the United States' targets, the *Paris Agreement* is very unlikely to put the world on a path to safe reductions of greenhouse gas emissions by 2100 without very aggressive post-2030 commitment (Rogelj et al. 2016). The United States targets to date reflect a futile attempt to negotiate, on the basis of political and socio-economic factors, with an ecological limit that is ecologically non-negotiable—in other words, they pose an unacceptable risk of catastrophe.

Second, only modest reductions in greenhouse gas emissions can be achieved using existing mechanisms under the *Clean Air Act*. Nordhaus (2007: 69) estimates that the *Clean Air Act* could be used to regulate sources responsible for only 70 to 80 percent of United States carbon dioxide emissions, or (because other greenhouse gases account for 20 percent of total greenhouse gas emissions) 59 to 67 percent of total greenhouse gas emissions. Moreover, a full 60 percent reduction in greenhouse gas emissions could be achieved only with complete elimination of the emissions subject to regulation under the Act. Many factors make this level of reduction virtually impossible—for example, the projected 57 percent increase in vehicle miles traveled in the United States from 2004 to 2030, which would require emissions per mile to decrease by 36 percent just to hold the total level of emissions steady (Nordhaus 2007: 70–71).

Third, Congressional efforts to move beyond the authority to regulate greenhouse gases in the *Clean Air Act* stalled after the United States House of Representatives passed the *American Clean Energy and Security Act of 2009* [*ACES Bill*].[34] The *ACES Bill* would have amended the *Clean Air Act* with a new Title on global warming pollution, with provisions that aimed to achieve a 20 percent reduction from 2005 levels of United States greenhouse gas emissions reduction from 2005 levels by 2050.[35] The massive bill would have implemented a complex cap-and-trade mechanism to attain those objectives, along with programs to increase energy efficiency and renewable energy use, promote "clean" transportation, improve the national electrical grid, support development of carbon capture and sequestration technologies, and support development of "clean energy" (including nuclear energy). The *ACES Bill* could not pass even with a President and majorities in both houses of Congress in favor of it in 2009. The prospects have worsened since, and in 2020 no clear path exists to new climate legislation that would put the United States on track to contribute to a global effort to keep greenhouse gas emissions below a cautious planetary boundary.

Last, even if the United States were to enact climate legislation, adequate implementation is far from certain. Almost 40 years after the *Clean Air Act* was adopted, National Ambient Air Quality Standards (NAAQSs) for more traditional pollutants have not been attained in many areas. Section 108 of the *Clean Air Act* provides that

> [f]or the purpose of establishing national primary and secondary ambient air quality standards, the Administrator shall within 30 days after December 31, 1970, publish, and shall from time to time thereafter revise, a list which includes each air pollutant—

> (A) emissions of which, in his judgment, cause or contribute to air pollution which may reasonably be anticipated to endanger public health or welfare; (B) the presence of which in the ambient air results from numerous or diverse mobile or stationary sources … .[36]

The EPA is required to establish these NAAQSs solely on the basis of the impacts of pollutants on human health, without accounting for technological or economic factors.[37]

To date, EPA has established NAAQSs for six pollutants: carbon monoxide, lead, nitrogen dioxide, two sizes of particulate matter (PM10 and PM2.5), ground-level ozone and sulfur dioxide.[38] The Act requires the States to identify "non-attainment areas" within their borders in which any of the NAAQSs are not attained and to develop plans for adjusting air emissions permits, revising motor vehicle programs or taking other measures in order to protect areas that are in attainment and to achieve attainment of NAAQSs in non-attainment areas. The persistent failure to attain NAAQSs despite these programs, particularly for ground-level ozone associated with motor vehicle emissions (Heinzerling 2005: 155), like the persistent failure to attain water quality standards through TMDLs, portends difficulty in achieving reductions in greenhouse gas emissions in the United States that would be needed to stay within ecologically-based system boundaries—even if adequate legislation were enacted.

Critical loads and levels of air pollutants in Europe

The critical loads and levels approach (Cresser 2000; Rockström et al. 2009: Supp at 6) was developed under the United Nations Economic Commission for Europe's *LRTAP*.[39] As with TMDLs and NAAQSs, the general concept behind this approach is to establish the threshold levels of ambient pollutants below which harmful effects on receptor ecosystems will be prevented. Critical load is defined as a "quantitative estimate of an exposure to one or more pollutants below which significant harmful effects on specified sensitive elements of the environment do not occur according to present knowledge" (CLRTAP 2015: I-1; Umweltbundesamt 2004: V-1). Critical levels are defined as "concentrations of pollutants in the atmosphere above which direct adverse effects on receptors, such as human beings, plants, ecosystems or materials, may occur according to present knowledge" (Umweltbundesamt 2004: III-1). Critical loads and levels are estimated using biological and other indicators of the systemic effects of pollutant exposure and loading. For example, critical levels with respect to vegetation are based in large part on the effects of ambient pollutants on "growth changes, yield losses, visible injury and reduced seed production" (Umweltbundesamt 2004: III-1) in five main vegetation categories: agricultural crops, horticultural crops, semi-natural vegetation, natural vegetation and forest trees.

Once critical loads and levels are estimated, they serve as a basis for limiting releases of pollutants to the environment from specific sources. The critical load and level approach uses complex modeling to estimate the emissions reductions and allocations that are needed to keep levels of individual pollutants or pollutants in combination below critical loads (pollutant deposition) and levels (ambient pollution) (CLRTAP 2015). Maps showing where critical levels and loads are exceeded (exceedance maps) are used to allocate the loads and levels to emissions from different countries; and maps of the concentrations and depositions of pollutants are used to assess the effects of pollutants in specific ecosystems (Umweltbundesamt 2004: II-1). Accounting for the dynamic nature systems in response to change, such as delayed biological effects of pollutant loading, the approach uses dynamic modeling to "assess time delays of recovery in regions where critical loads cease being exceeded and time delays of damage in regions where critical loads continue to be exceeded" (Umweltbundesamt 2004: VI-1). Although dynamic modeling has been applied successfully with respect to pollutants like sulfur dioxide that cause acidification, it has been less successful with respect to the more complex biogeochemical processes and feedbacks involved in eutrophication (Spranger et al. 2008: 484).

Despite the complexities and uncertainties that must be addressed, reasonable progress has been made in estimating critical loads and levels. It has been used to establish critical loads for eutrophication and acidification due to sulfur and nitrogen-based pollution and for heavy metals, specifically cadmium, lead and mercury; and to develop critical levels for ozone, nitrogen oxides, sulfur dioxide and ammonia (Umweltbundesamt 2004: I-5). As well, a methodology exists to examine the combined effects of multiple pollutants on ecosystems through "critical load functions" (Spranger et al. 2008: 483).

Exceedances of critical loads and levels are still widespread in Europe but are decreasing. As with NAAQSs and TMDLs in the United States, technical and economic factors determine the extent and rate of progress. The initial levels of emissions reductions required under the *LRTAP* were based solely on technical and economic considerations, such as best available technologies, without taking into account the connection between exceedances and effects on receptor ecosystems (Umweltbundesamt 2004: I-3). In the next set of protocols, which took effect in 1994,

> cutting sulphur dioxide emissions to achieve deposition levels below critical loads was not feasible for all ecosystems in Europe. Even so, the negotiations [of abatement measures] were based on the assessment of environmental effects and the protection of ecosystems as well as technical and economic considerations.
>
> (Umweltbundesamt 2004: I-3)

Gradual attainment of critical loads and levels is a long-term objective of the eight protocols that set out the obligations of the signatories to reduce emissions, but complete elimination of exceedances is not necessarily anticipated.[40] The European Union's 2016 *National Emission Ceilings Directive* acknowledges that technical feasibility and cost effectiveness take precedence over full achievement of critical loads and levels through 2030 and beyond. Thus, for example, implementation of the Directive may leave about 67 percent of ecosystems within the European Union and the United Kingdom in exceedance of critical loads for eutrophication (Hettelingh and Posch 2019). Likewise, a 2008 study concluded:

> The eutrophication of terrestrial ecosystems due to nitrogen emissions into the atmosphere in Europe will remain, and poses a risk to biodiversity and other environmental targets on a continental scale. Animal husbandry increasingly determines magnitude and spatial pattern of critical load exceedance, i.e. the risk to sustainability with respect to the long-range transport of air pollutants. This conclusion holds for eutrophication as well as acidification.
>
> (Spranger et al. 2008: 485–486)

The *LRTAP* has been praised as

> a successful international mechanism for alleviating the global problem of air pollution [that has resulted in] a significant production of data concerning air pollution, reduction of a number of pollutants, an increasing exchange of scientific information, and a stimulation of national pollution reduction efforts.
>
> (Shapiro 2000: 216)

By contrast, the effectiveness of the *LRTAP* has been questioned because political and economic considerations have consistently taken precedence over the Convention's environmental goals (Pelletier 2010: 221). Without question, although it sets limits using rigorous science regarding impacts on receptor ecosystems, the critical loads and levels approach under the *LRTAP* gives economic, technical and political concerns precedence over ecological limits.

Cap-and-trade program for sulfur dioxide under the United States *Clean Air Act*

Another only marginally effective mechanism that bases restrictions on an activity or pollution on ecosystem effects is the cap-and-trade program under the United States *Clean Air Act* to control acid rain due to sulfur dioxide emissions.[41] In amendments to the *Clean Air Act* in 1990, a cap-and-trade mechanism was established for sulfur dioxide emissions from electric utilities and other air pollution sources. The program satisfies some of the United States' obligations under the *LRTAP.* After establishing a gradually diminishing cap on emissions from the entire category of sources, an initial set of quotas of emissions was allocated to the sources so as to respect the initial cap, and a market was created to allow sources to buy and sell the right to emit sulfur dioxide within the overall cap. The program has two phases, the initial phase covering the largest electric utilities and the second phase, which began in 2000, bringing smaller sources into the emissions market.[42]

A key driver of this legislation was the desire to reduce acid rain in areas downwind of major sulfur dioxide sources, particularly coal-fired power plants. In enacting the new provisions, the United States Congress found that acid rain is a problem of national and international significance that "represents a threat to natural resources, ecosystems, materials, visibility, and public health."[43] The purpose of the Acid Rain Program is

> to reduce the adverse effects of acid deposition through reductions in annual emissions of sulfur dioxide of ten million tons from 1980 emission levels, and, in combination with other provisions of this Act, of nitrogen oxides emissions of approximately two million tons from 1980 emission levels, in the forty-eight contiguous States and the District of Columbia.[44]

Nothing in the Act links the desired reductions to a particular objective for the ecosystems to which acid rain causes harm, even if some amount of protection of those ecosystems was anticipated. Unlike Europe's critical loads and levels approach, it does not establish ecologically determined limits. After reviewing the legislative process leading to adoption of the 1990 amendments, Heinzerling (1995: 303) concluded that "Congress appears to have paid scarcely any attention to the pollution level set by the … Amendments and to have concentrated instead on satisfying powerful interest groups through its allocation of permits." Even greater reductions in sulfur dioxide emissions might have been achieved by allowing command-and-control provisions that the amendment effectively repealed to remain in effect, albeit at greater cost to industry (Moore 1992: 40–42).

The Acid Rain Program in the United States is widely viewed as a success. According to one source, "[e]missions of SO2 have been reduced at a faster rate and at considerably lower cost than expected [and] by 2005, SO2 emissions from the power plants included in the program had fallen 35% from 1990 levels" (Doremus and Hanemann 2008: 802). However, some of the decline of sulfur dioxide and nitrogen oxides emissions is due to an increase in the availability of relatively cheap low-sulfur coal in the 1990s (Moore 1992). Moreover, the European Union achieved a 70 percent reduction in sulfur dioxide emissions between 1980 and 2001, almost twice the percentage reduction achieved in the United States over the same time period: in 2001 the emissions of sulfur dioxide per unit of GDP were 494 kilograms per million euros in Europe and 1,150 kilograms per million euros in the United States (European Commission 2004: 8).

The Acid Rain Program has also had only partial success in reducing impacts on downwind ecosystems. A 2005 report on the effects of acid rain on lakes in the Adirondack Mountains in New York State states:

> Despite many denials, acid rain did and does exist and was and is a significant ecological threat. It has been significantly but only partially abated, and as we write this in 2005 the Adirondacks are still threatened. More

> abatement is needed, and even then the recovery will be slow. Acid rain … is cumulative and persistent.
>
> (Jenkins et al. 2005: 1)

According to a 2001 study, "power plants must cut SO2 emissions an additional 80% in order to allow sensitive waters and soils in the Northeast to recover" (Doremus and Hanemann 2008: 802). In short, the Acid Rain Program may be a success in terms of its ability to achieve a certain amount of pollution reduction with lower costs to industry than would have been the case with traditional command-and-control standards, but not in terms of reducing pollution to a sufficient degree to allow ecosystems damaged by acid rain to return to their pre-disturbance condition.

Tradeable pollution rights mechanisms like the Acid Rain Program in the United States are suitable for some environmental problems but not others. First, while they may work well for substances like sulfur dioxide and carbon dioxide, where the sources and environmental effects are relatively fungible, they are less appropriate in the case of the exchange of wetlands or other ecosystems, where no such fungibility among what is being traded exists (Salzman and Ruhl 2000). In the case of toxic pollutants like mercury, which pose the problem of toxic hotspots that trading alone will not sufficiently address, trading is virtually never appropriate (Doremus and Hanemann 2008: 803).

A relatively successful system-based limit: *The Montreal Protocol on Substances that Deplete the Ozone Layer*

One moderately successful regulation that imposes limits on pollutants or activity in order to avoid unacceptable harm to ecological or biogeochemical systems is the international regime for controlling substances that deplete stratospheric ozone. Although this example may provide some basis for imagining a broader, more comprehensive regime of systems-based governance mechanisms that would constitute ecological law, it is in many respects a special case.

The *Montreal Protocol on Substances that Deplete the Ozone Layer* [*Montreal Protocol*][45] was adopted in 1987, two years after the ozone hole over Antarctica was discovered. The agreement sets forth a mandatory schedule of reductions of the production and consumption of substances that deplete stratospheric ozone, along with a process for conducting periodic scientific assessments of the control measures with a view to adjusting them as necessary.[46] The signatory Parties have regularly tightened the initial set of control measures in a series of adjustments and amendments, both by decreasing the allowable production and consumption of ozone-depleting substances and by bringing additional ozone-depleting substances under the Protocol. The Parties have agreed through those adjustments to the phase-out of the consumption and production of chlorofluorocarbons [CFCs], halons, other fully halogenated CFCs, carbon tetrachloride, methyl chloroform and methyl bromide, with exceptions for "essential uses" and attenuated control schedules for developing countries.[47] For example, provisions were adopted in 2007 to phase out ozone-depleting hydrochlorofluorocarbons (HCFCs) and in 2016 to phase out hydrofluorocarbons (HFCs), both of which had served as interim replacements for more destructive substances banned earlier (Roberts 2009: 35; Albrecht and Parker 2019).

By 1999, the *Montreal Protocol* had resulted in an 85 percent global decrease in the production and consumption of ozone-depleting substances (UNEP Ozone Secretariat 1999). By 2007, ozone-depleting substances had declined by 95 percent (Roberts 2009: 33), and ozone-depleting substances were in overall decline in both the troposphere and the stratosphere following a peak in stratospheric levels in the late 1990s (UNEP Ozone Secretariat 2007: 4). The Ozone Secretariat predicted in 2007 that if compliance with the Protocol continued, the levels of ozone in the stratosphere would return to their pre-1980 levels, when no hole in the ozone layer existed, by 2050 at mid latitudes and a decade or two later at the poles (UNEP Ozone Secretariat 2007). A 2019 study confirmed that healing of the Antarctic ozone hole, in which ozone levels begin to rise after having declined and stabilized, is underway (Solomon et al. 2016). Planetary boundaries researchers consider stratospheric ozone to be "a good example where concerted human effort and wise decision making seem to have enabled us to stay within a planetary boundary" (Rockström et al. 2009: 15).

The *Montreal Protocol* is widely seen as an environmental success story (Sunstein 2007: 73; Roberts 2009: 33; Winchester 2009: 11; Esty 2009: 427; Rockström et al. 2009: 15), despite some concern that it might be ignoring some significant stockpiles of ozone-depleting substances (Roberts 2009: 33, 35) and that global heating may impede recovery of the ozone layer (Reeves 2005: 54–55). However, relatively, its success is not likely to be repeated with other global problems, especially economy-wide ones like climate change. Victor (2008: 76) explains:

> Several factors led to this unusually rapid response from the international community to the problem of CFCs once it was discovered. These included the scientific consensus on the chemistry of CFCs in the stratosphere, heightened public awareness, the availability of profitable and less harmful substitute chemicals such as HFCs and HCFCs (which are also potent greenhouse gases), and the small number of producers, making regulations easy to enforce, though a black market in CFCs developed undermining for a time the new regulations.

In light of these mostly favorable factors, in key countries like the United States, the societal benefits were seen to be much higher than the societal costs, which is unlikely to be the case for climate change (Sunstein 2007: 73–75). Although scientific consensus on the role of CFCs and other substances in depleting the stratospheric ozone layer was quite strong, it was not absolute, and the *Montreal Protocol* therefore can be seen as a relatively successful application of the precautionary principle to avoid catastrophe for human health and the global environment (Sunstein 2007: 73).

The *Montreal Protocol* has been successful in ways that TMDLs, NAAQSs, control of greenhouse gases in the United States and the critical levels and loads approach in Europe have not. Specifically, although at least some of those other mechanisms have identified thresholds on the tolerance of receptor systems to the effects of pollutants, they have not yielded effective controls on the industrial and other processes that cause those effects. By contrast, the prevailing view appears to be that the *Montreal Protocol* is on a predictable track to eliminating ozone-depleting substances so as to return the stratospheric ozone layer to its pre-disturbance condition within this century. However, that the *Montreal Protocol* was relatively successful only because of a unique set of circumstances highlights how more generally, contemporary environmental law is not up to the task of preventing human activities from transgressing critical global ecological limits. In Part III, I examine how reorienting society and law toward a mutually enhancing human-Earth relationship presents enormous challenges, but also a more hopeful prospect for humans and Earth's other living beings.

Notes

1 33 U.S.C. §§ 1251 *et seq.* (2006).
2 42 U.S.C. §§ 7401 *et seq.* (2006).
3 *Directive of the European Parliament and of the Council 2001/81/EC of 23 October 2001 on national emission ceilings for certain atmospheric pollutants* [2001] O.J. L 309/22.
4 13 November 1979, 1302 U.N.T.S. 217 (entered into force 16 March 1983).
5 16 September 1987, 1522 U.N.T.S. 3 (entered into force 1 January 1989).
6 *Clean Water Act* § 303(d)(1)(A).
7 *Clean Water Act* § 303(d)(1)(C). Section 304(a)(2) provides that "the [EPA] Administrator, after consultation with appropriate Federal and State agencies and other interested persons, shall develop and publish, within one year after October 18, 1972 (and from time to time thereafter revise) information … for the purpose of section 1313 of this title, on and the identification of pollutants suitable for maximum daily load measurement correlated with the achievement of water quality objectives."
8 40 C.F.R. § 130.2(g), (h) and (i).
9 40 C.F.R. § 130.7(c)(1).
10 *Clean Water Act* §§ 303(d)(4), 303(e), 33 U.S.C. §§ 1313(d)(4), 1313(e); 40 C.F.R. § 130.5.
11 *Clean Water Act* § 303(c)(2)(A).
12 *Clean Water Act* § 304(a)(1), 33 U.S.C. § 1314(a)(1).
13 U.S. Environmental Protection Agency, *National Water Quality Inventory: Report to Congress, 2004 Reporting Cycle*, at 2.
14 EPA, Impaired Waters and TMDLs, www.epa.gov/tmdl/impaired-waters-restoration-process.
15 EPA, *Withdrawal of Revisions to the Water Quality Planning and Management Regulation and Revisions to the National Pollutant Discharge Elimination System Program in Support of Revisions to the Water Quality Planning and Management Regulation*, 68 FR 13608 (March 19, 2003).
16 EPA, *National Summary of Impaired Waters and TMDL Information*, https://iaspub.epa.gov/waters10/attains_nation_cy.control?p_report_type=T
17 EPA, Impaired Waters and Mercury, www.epa.gov/tmdl/impaired-waters-and-mercury.

18 *Massachusetts v. EPA*, 549 U.S. 497 (2007).
19 *Clean Air Act* § 202(a)(1), 42 U.S.C. §7521(a)(1) (2006).
20 *Clean Air Act* § 302(h), 42 U.S.C. § 7602(g) (2006).
21 549 U.S. at 528–529.
22 *Clean Air Act* § 302(h), 42 U.S.C. § 7602(h) (2006).
23 549 U.S. at 533.
24 549 U.S. at note 26. Many pollutants with local effects, like sulfur dioxide and mercury, also have broader systemic effects (such as acid rain and contamination of fish from mercury deposition into fish-bearing waters) that the EPA has taken into account in regulating them.
25 U.S. Environmental Protection Agency, *Endangerment and Cause or Contribute Findings for Greenhouse Gases Under Section 202(a) of the Clean Air Act*, 74 Fed. Reg. 66496 (2009) (codified at 40 C.F.R. Chapter 1) [*GHG Endangerment Finding*].
26 U.S. EPA and NHTSA, *Light-Duty Vehicle Greenhouse Gas Emission Standards and Corporate Average Fuel Economy Standards; Final Rule*, 75 Fed. Reg. 25324 (2010) (codified at 40 CFR Parts 85, 86 and 600; 49 CFR Parts 531, 533, 536, et al.) [*Light-Duty Vehicle Rule*].
27 U.S. EPA and NHTSA, *2017 and Later Model Year Light-Duty Greenhouse Gas Emissions and Corporate Average Fuel Economy Standards; Final Rule,* 77 Fed. Reg. 62624 (2012) (codified at 40 CFR Parts 85, 86 and 600 and 49 CFR Parts 523, 531, 533, 536 and 537).
28 U.S. EPA and NHTSA, *The Safer Affordable Fuel-Efficient (SAFE) Vehicles Rule for 2021–2026: Passenger Cars and Light Trucks; Final Rule* (2020) (to be codified at 40 CFR Parts 85 and 86 and 49 CFR Parts 523, 531, 533, 536 and 537) [online]: www.epa.gov/sites/production/files/2020-03/documents/final-fr-safe-preamble-033020.pdf.
29 U.S. EPA, *Carbon Pollution Emission Guidelines for Existing Stationary Sources: Electric Utility Generating Units; Final Rule* (October 23, 2015), 80 Fed. Reg. 64662 et seq (codified at 40 CFR Part 60).
30 U.S. EPA, *Repeal of the Clean Power Plan; Emission Guidelines for Greenhouse Gas Emissions From Existing Electric Utility Generating Units; Revisions to Emission Guidelines Implementing Regulations; Final Rule* (July 8, 2019), 84 Fed. Reg. 32520 et seq (codified at 40 CFR Part 60).
31 9 May 1992, 1771 U.N.T.S. 107 (entered into force 21 March 1994).
32 U.S. Department of State, *On the U.S. Withdrawal from the Paris Agreement,* Press Statement of Secretary of State Pompeo (November 4, 2019) [online]: www.state.gov/on-the-u-s-withdrawal-from-the-paris-agreement/.
33 The *Paris Agreement* aims to hold "the increase in the global average temperature to well below 2 °C above pre-industrial levels and pursuing efforts to limit the temperature increase to 1.5 °C above pre-industrial levels, recognizing that this would significantly reduce the risks and impacts of climate change." Article 2(1)(a).
34 Bill H.R. 2454, *American Clean Energy and Security Act of 2009*, 111th Cong., 2009 [*ACES Bill*] (not enacted).
35 *ACES Bill* § 311 (proposed *Clean Air Act* § 702).
36 *Clean Air Act* § 108(a)(1), 42 U.S.C. § 7408(1)(1) (2006).
37 *Clean Air Act* § 109(b)(1), 42 U.S.C. § 7409(b)(1) (2006).
38 U.S. Environmental Protection Agency, *National Ambient Air Quality Standards (NAAQS)* (2010) [online]: www.epa.gov/air/criteria.html.
39 Fifty-one countries, including the European Union, nearly all European States, Russia, Canada and the United States, have ratified the Convention. United Nations [online]: https://treaties.un.org/Pages/ViewDetails.aspx?src=TREATY&mtdsg_no=XXVII-1&chapter=27&clang=_en.
40 See, e.g. *Protocol to the 1979 Convention on Long-range Transboundary Air Pollution to Abate Acidification, Eutrophication and Ground-level Ozone* (30 November 1999). Document of the Economic and Social Council EB.AIR/1999/1 (entered into

force 17 May 2005) [*1999 Gothenburg Protocol*] Article 2; EU, *Directive 2016/2284 of the European Parliament and of the Council of 14 December 2016 on the reduction of national emissions of certain atmospheric pollutants [National Emissions Ceiling Directive]*, amending Directive 2003/35/EC and repealing Directive 2001/81/EC, OJ L 344, 17.12.2016, pp. 1–31.

41 *Clean Air Act* Title IV, 42 U.S.C. §§ 7651–7651o (2006).

42 *Clean Air Act* § 405, 42 U.S.C. § 7651d (2006).

43 *Clean Air Act* § 401(a), 42. U.S.C. § 7651(a) (2006).

44 *Clean Air Act* § 401(b), 42 U.S.C. § 7651(b) (2006).

45 16 September 1987, 1522 U.N.T.S. 3 (entered into force 1 January 1989).

46 *Montreal Protocol*, arts. 2, 6, 11.

47 *Adjustments to the Montreal Protocol agreed by the Second, Fourth, Seventh, Ninth, Eleventh and Nineteenth Meetings of the Parties*; *London Amendment to the Montreal Protocol,* 29 June 1990 (entered into force 10 August 1992); *Copenhagen Amendment to the Montreal Protocol*, 25 November 1992 (entered into force 14 June 1994); *Montreal Amendment to the Montreal Protocol*, 17 September 1997 (entered into force 10 November 1999); *Beijing Amendment to the Montreal Protocol,* 3 December 1999 (entered into force 12 February 2002); *Kigali Amendment to the Montreal Protocol,* 15 October 2016 (entered into force 1 January 2019).

References

Adler, R.W. 2009. Freshwater: sustaining use by protecting ecosystems. *Environmental Law Reporter News & Analysis* 39: 10309–10315.

Albrecht, F., and C.F. Parker. 2019. Healing the Ozone Layer: The Montreal Protocol and the Lessons and Limits of a Global Governance Success Story, in 't Hart, P. and M. Compton, eds., *Great Policy Successes.* New York: Oxford University Press.

Boyd, J. 2000. *The New Face of the Clean Water Act: A Critical Review of EPA's Proposed TMDL Rule.* Resources for the Future, Discussion Paper 00–12.

Carbon Action Tracker. 2020. *USA* [online]: https://climateactiontracker.org/countries/usa/sources/

Caudill, D.S., and D.E. Curley. 2008. Strategic idealizations of science to oppose environmental regulation: a case study of five TMDL controversies. *University of Kansas Law Review* 57: 251–312.

CLRTAP. 2015. *Introduction, Chapter 1 of Manual on methodologies and criteria for modelling and mapping critical loads and levels and air pollution effects, risks and trends.* UNECE Convention on Long-range Transboundary Air Pollution.

Cresser, M.S. 2000. The critical loads concept: milestone or millstone for the new millennium? *The Science of the Total Environment* 249: 51–62.

Doremus, H., and W.M. Hanemann. 2008. Of babies and bathwater: why the Clean Air Act's cooperative federalism framework is useful for addressing global warming. *Arizona Law Review* 50: 799–834.

Esty, D.C. 2009. Revitalizing global environmental governance for climate change. *Global Governance* 15: 427–434.

European Commission. 2004. Comparison of the EU and US Approaches Towards Acidification, Eutrophication and Ground Level Ozone in *Assessment of the Effectiveness of European Air Quality Policies and Measures*, Doc. B4–3040/2003/365967/MAR/C1 [online]: http://ec.europa.eu/environment/archives/cafe/activities/pdf/case_study1.pdf.

Gupta, S., D.A. Tirpak, N. Burger, J. Gupta, N. Höhne, A.I. Boncheva, G.M. Kanoan, C. Kolstad, J.A. Kruger, A. Michaelowa, S. Murase, J. Pershing, T. Saijo and A. Sari.

2007. Policies, Instruments and Co-operative Arrangements, in B. Metz, O.R. Davidson, P.R. Bosch, R. Dave and L.A. Meyer (eds), *Climate Change 2007: Mitigation. Contribution of Working Group III to the Fourth Assessment Report of the Intergovernmental Panel on Climate Change*. New York: Cambridge University Press.

Heinzerling, L. 1995. Selling pollution, forcing democracy. *Stanford Environmental Law Journal* 14: 300–344.

Heinzerling, L. 2005. Risking it all. *Alabama Law Review* 57: 103–117.

Hettelingh, J.P., and M. Posch. 2019. Critical load exceedances under equitable nitrogen emission reductions in the EU28. *Atmospheric Environment* 211: 113–119.

Jenkins, J., K. Roy, C. Driscoll and C. Buerkett. 2005. *Acid Rain and the Adirondacks: A Research Summary*. Ray Brook, NY: Adirondacks Lakes Survey Corporation.

Karr, J.R., and C.O. Yoder. 2004. Biological assessment and criteria improve total maximum daily load decision making. *Journal of Environmental Engineering* June: 594–604.

Keyes, A.T., K.L. Lambert, D. Burtraw, J.J. Buonocore, J.I. Levy and C.T. Driscoll. 2019. The affordable clean energy rule and the impact of emissions rebound on carbon dioxide and criteria air pollutant emissions. *Environmental Research Letters* 14: 044018.

Moore, C.A. 1992. The 1990 Clean Air Act amendments: silk purse or sow's ear? *Duke Environmental Law & Policy Forum* 2: 26–58.

Nordhaus, W. 2007. *The Challenge of Global Warming: Economic Models and Environmental Policy*. New Haven, CT: William Nordhaus.

Pelletier, N. 2010. Of laws and limits: An ecological economic perspective on redressing the failure of contemporary global environmental governance. *Global Environmental Change* 20: 220–228.

Percival, R.V. 2007. Environmental law in the twenty-first century. *Virginia Environmental Law Journal* 25: 1–35.

Reeves, H. 2005. *Mal de terre*. Paris: Éditions du Seuil.

Roberts, M.W. 2009. The Montreal Protocol must act to prevent global climate change while restoring the ozone layer. *Sustainable Development Law & Policy* 9: 33–40, 66–67.

Rockström, J., W. Steffen, K. Noone, Å. Persson, F.S. Chapin III, E. Lambin, T.M. Lenton, M. Scheffer, C. Folke, H. Schellnhuber, B. Nykvist, C.A. De Wit, T. Hughes, S. van der Leeuw, H. Rodhe, S. Sörlin, P.K. Snyder, R. Costanza, U. Svedin, M. Falkenmark, L. Karlberg, R.W. Corell, V.J. Fabry, J. Hansen, B. Walker, D. Liverman, K. Richardson, P. Crutzen and J. Foley. 2009. Planetary boundaries: exploring the safe operating space for humanity. *Ecology and Society* 14(2): 32 [online] URL: www.ecologyandsociety.org/vol.14/iss2/art32/

Rogelj, J., M. den Elzen, N. Höhne, T. Fransen, H. Fekete, H. Winkler, R. Schaeffer, F. Sha, K. Riahl and M. Meinshausen. 2016. Paris Agreement climate proposals need a boost to keep warming well below 2°C. *Nature* 534: 631–639.

Salzman, J., and J.B. Ruhl. 2000. Currencies and the commodification of environmental law. *Stanford Law Journal* 53: 607–694.

Shapiro, W.J. 2000. Protocol to abate acidification, eutrophication and ground level ozone. *Colorado Journal of International Environmental Law & Policy* 11(1999 Yearbook): 208–219.

Solomon, S., D.J. Ivy, D. Kinnison, M.J. Mills, R.R. Neely III and A. Schmidt. 2016. Emergence of healing in the Antarctic ozone layer. *Science* 353(6296): 269–274. Spranger, T., J-P. Hettelingh, J. Slootweg and M. Posch. 2008. Modelling and mapping long-term risks due to reactive nitrogen effects: an overview of LRTAP convention activities. *Environmental Pollution* 154: 482–487.

Sunstein, C.R. 2007. *Worst-Case Scenarios.* Cambridge MA: Harvard University Press.

Umweltbundesamt. 2004. *Manual on Methodologies and Criteria for Modelling and Mapping Critical Loads & Levels and Air Pollution Effects, Risks and Trends.* Berlin: Umweltbundesamt.

United Nations Environment Programme, Ozone Secretariat. 1999. *Report of the Eleventh Meeting of the Parties to the Montreal Protocol on Substances that Deplete the Ozone Layer* [online]: ozone.unep.org/Meeting_Documents/mop/11mop/11mop-10.e.pdf.

United Nations Environment Programme, Ozone Secretariat. 2007. *Synthesis Report of the 2006 Assessments of the Scientific Assessment Panel, the Environmental Effects Assessment Panel and the Technology and Economic Assessment Panel*, No. UNEP/OzL.Pro.WG.1/27/3. Nairobi: United Nations Environment Program.

USEPA. 2017. *National Water Quality Inventory: Report to Congress.* Document No. EPA 841-R-16-011. Washington: USEPA.

Victor, P. 2008. *Managing Without Growth.* Cheltenham, UK: Edward Elgar.

Winchester, N.B. 2009. Emerging global environmental governance. *Indiana Journal of Global Legal Studies* 16(1): 7–24.

Part III

Ecological law

Systems-based rules for a flourishing Earth

7 What is ecological law?

The Anthropocene prompts re-examination of the conceptual foundations of normative social constructs such as law. In Parts I and II, I presented an irreconcilable clash between the currently dominant growth-insistent narrative and a narrative that gives primacy to containing the human enterprise within ecological limits in a life-enhancing way. The growth-insistent narrative relies on strong notions of private property rights and State sovereignty, which have contributed significantly to the rise of the human-Earth dilemma in the Anthropocene. In Part III, I turn to the need to shift the foundations of law in the Anthropocene away from the growth-insistent narrative and towards a limits-insistent narrative—in other words, toward ecological law.

So, what is ecological law? Until recently, the rare uses of the term "ecological law," at least in the English language, did not entirely capture the meaning intended in this book. Howarth (2006: 3–4) makes a distinction between environmental quality law, which he defines as the law controlling the quality of environmental media such as air and water, and ecological law, which he defines as the law related to the relationships between environmental media and their living components. Kolbasov (1989), whose use of the term dates back to 1976, takes the opposite tack. Noting the fragmentation in the Soviet Union of legal regimes related to the interaction of society and nature, he describes ecological law as including "the entire complex of issues of nature-use and nature protection, excluding those that pertain to constitutional, administrative, state, civil, economic, labor and criminal law" (Kolbasov 1989: 77). Thus, Kolbasov would combine under the rubric of ecological law all legal regimes dealing with the use of land, minerals, forests and water with those dealing with the protection of wildlife and environmental media like air and water, but not if they are covered by the various other legal orders he lists.

As used in this book, ecological law comprises and transcends both of Howarth's concepts and it goes well beyond Kolbasov's constrained attempt to harness together fragmented legal regimes related to the human-Earth relationship. It is more akin to the notion of "ökologischen Recht" (ecological law) in German legal scholarship as early as the 1980s (Bosselmann 1986: 3). The German understanding of ecological law comports with the ecologically imbued notion of law of Berry (1999: 61):

> To achieve a viable human-Earth situation a new jurisprudence must envisage its primary task as that of articulating the conditions for the integral functioning of the Earth process, with special reference to a mutually enhancing human-Earth relationship. Within this context the various components of the Earth—the land, the water, the air, and the complex of life systems—would each be a commons. Together they would constitute the integral expression of the Great Commons of the planet Earth to be shared in proportion to need among all members of the Earth Community.

In Berry's view of the law, a sustainable human-Earth relationship requires that

> inherent rights of the natural world [be] recognized as having legal status [and t]he entire question of possession and use of the Earth, either by individuals or by establishments, needs to be considered in a more profound manner than Western society has ever done previously.
>
> (Berry 1999: 61)

For Berry, "[e]cology is not a part of law; law is an extension of ecology" (Berry 1999: 84).

More recent reflections of Berry's ecological conception of law as a guide for a mutually enhancing human-Earth relationship are in Bosselmann's "principle of sustainability" (Bosselmann 2008) and Boyd's similar notion of "sustainability law" (Boyd 2004). Boyd (2004: 364–365) explains:

> Sustainability law would not be about merely *mitigating* the damage inflicted by industrial economies and western lifestyles. Sustainability law would focus on *transforming* the relationship between humans and the natural environment from one based on minimizing harm to one based on maximizing harmony. Instead of asking if we can limit the ecological damage caused by contemporary industrial society, sustainability law asks if we can do things in a completely different way that avoids creating environmental problems in the first place. Sustainability law will challenge the belief that human activities must inevitably damage the natural world.

Sustainability law "would be firmly rooted in science and the laws of nature, beginning with a clear understanding of the laws of thermodynamics and explicit recognition of the biophysical limits of the planet Earth" (Boyd 2004: 367). Unlike Kolbasov's notion of ecological law, sustainability law would be embedded within and throughout the legal, social and economic infrastructure, not a specialty area of the law (Boyd 2004: 364).

Bosselmann (2008: 1–2, 53) rejects "weak sustainability" that merely integrates economic growth with social prosperity and environmental protection and insists that the core of sustainability is about respecting and maintaining the Earth's ecological integrity. His central thesis is that with a grounding in ecological integrity, "sustainability has the historical, conceptual and ethical quality typical for a fundamental principle of law" (Bosselmann 2008: 4). Bosselmann's principle of sustainability is as broad and fundamental as foundational principles like equality, justice and freedom, establishing a paramount "duty to protect and restore the integrity of the Earth's ecological systems" (Bosselmann 2008: 53). In his view, the Earth Charter, with its emphasis on the interdependence of all life forms, the value of all living things regardless of their value to humans, and the dignity and potential of human beings, is the most profound expression of the broad and foundational legal principle of sustainability (Bosselmann 2008: 57, 73–75).

Ecological law has much in common with Boyd's sustainability law and Bosselmann's principle of sustainability. However, the term "ecological law" makes clear that global ecological limits have primacy over the economic and social spheres that are usually associated, along with environmental concerns, with "sustainability"—often in a confusing way, or in a way that gives primacy to economic growth (Boer 2000; Pelletier 2010: 224; Garver 2013). Both Boyd and Bosselmann recognize this essential primacy of ecological limits and integrity.

Ecological law is firmly grounded in the science of how the Earth works and of complex systemic thresholds in the global ecosystem, and therefore combines legal principles with scientific laws of ecology as expressed, for example, in planetary boundaries. Yet, science alone cannot determine its elements. The laws of thermodynamics and the science of ecology, both central to the scientific understandings on which ecological economics and ecological law are built, cannot simply be plugged in to make the human-Earth relationship right. Both a thriving tropical forest and a lifeless toxic waste dump have their own ecology, and both obey the laws of thermodynamics. The same holds true for an Earth with or without human survival. Scientific laws and principles apply to whatever circumstances the Earth and its inhabitants and the cosmos offer up. But ecological law is not indifferent to those circumstances, or to the human place in them. Beyond science, ecological law derives from a fundamental normative choice to manage the human enterprise so as to offer up circumstances to which ecology, thermodynamics and the other sciences will apply in ways that preserve ecological integrity and keep the human-Earth relationship flourishing. The planetary boundaries and other ecological indicators help describe those circumstances, and they provide a basis for the normative decision inherent in ecological law, as explained further in Chapter 10.

Ecological law is open to a pluralistic notion of law, so as to include normative regimes that transcend traditional notions of law but are still successful in conforming behavior to desired objectives. As Bosselmann (2008: 53–54, 62–67) notes, ecological law (or the principle of sustainability) still lacks the broad, normative impact of a widely accepted guiding legal principle, even if it has increasingly influenced the legal sphere. A driving motivation behind the 2016 *Oslo Manifesto for Ecological Law and Governance*[1] was to harness the ideas to date on ecological approaches to law, give ecological law a higher profile among legal scholars and practitioners and establish a process for ongoing refinement of ecological law. A key part of this process was the creation of the

Ecological Law and Governance Association (ELGA). The manifesto describes the foundations of ecological law as follows:

> The ecological approach to law is based on ecocentrism, holism, and intra-/intergenerational and interspecies justice. From this perspective, or world-view, the law will recognise ecological interdependencies and no longer favour humans over nature and individual rights over collective responsibilities. Essentially, ecological law internalizes the natural living conditions of human existence and makes them the basis of all law, including constitutions, human rights, property rights, corporate rights and state sovereignty.

Note

1 Ecological Law and Governance Association (ELGA), Founding Documents [online]: www.elga.world/oslo-manifesto.

References

Berry, T. 1999. *The Great Work: Our Way Into the Future*. New York: Three Rivers Press.

Boer, B. 2000. Sustainability law for the new millennium and the role of environmental legal education. *Water, Air, and Soil Pollution* 123: 447–465.

Bosselmann, K. 1986. Eigene rechte für die natur? Ansätze einer ökologischen rechtsauffassung. *Kritische Justiz* 19(1): 1–22.

Bosselmann, K. 2008. *The Principle of Sustainability: Transforming Law and Governance*. Burlington, VT: Ashgate Publishing Co.

Boyd, D.R. 2004. Sustainability law: (R)Evolutionary directions for the future of environmental law. *Journal of Environmental Law and Practice* 14: 357–384.

Garver, G. 2013. The rule of ecological law: The legal complement to degrowth economics. *Sustainability* 5(1): 316–337 [online]: www.mdpi.com/2071-1050/5/1/316/htm

Howarth, W. 2006. The progression towards ecological quality standards. *Journal of Environmental Law* 18(1): 3–35.

Kolbasov, O.S. 1989. The concept of ecological law. *Connecticut Journal of International Law* 4: 267–278.

Pelletier, N. 2010. Of laws and limits: an ecological economic perspective on redressing the failure of contemporary global environmental governance. *Global Environmental Change* 20: 220–228.

8 A mutually enhancing human-Earth relationship as the primary goal for law

The conceptual foundations of a limits-insistent narrative for the human prospect, and of ecological law, are engrained in Thomas Berry's powerful invocation of a mutually enhancing human-Earth relationship (Berry 1999). A mutually enhancing human-Earth relationship is one in which humans individually and collectively see themselves as members, not masters, of the entire community of life on Earth and interact with Earth and the life it supports respectfully and "for the benefit of the larger community as well as ourselves" (Berry 1999: 5).

The focus on limits inherent in a mutually enhancing human-Earth relationship has roots in Malthus's warnings about population growth overtaking food production (Sieferle 2001) and in John Stuart Mill's vision of a steady-state economy (Alexander 2011; Daly 2015). It is also reflected in more recent notions, such as the planetary boundaries concept (Rockström et al. 2009), which is discussed in detail in Chapter 10. The planetary boundaries provide a broad set of global and sub-global measures with which to assess key aspects of whether the human-Earth relationship is mutually enhancing.

With respect to law, the notion of a mutually enhancing human-Earth relationship inspired the concepts of Earth Jurisprudence and Wild Law (Cullinan 2011), which legal scholars continue to elaborate and develop (Burdon 2011; Burdon 2015). Other concepts or conceptual frameworks that are relevant to the notion of a mutually enhancing human-Earth relationship include ecological integrity, "right relationship" (Brown and Garver 2009), deep ecology, strong sustainability, ecological economics, degrowth and "safe and just operating space for humanity" (Rockström et al. 2009; Raworth 2017; D'Alisa et al. 2015; Steffen et al. 2015).

These broad framings require critical analysis and refinement if they are to serve as practical and coherent guides for implementing a systems approach for moving the conceptual constructs of law and governance toward ecological law. As well, terms relevant to a mutually enhancing human-Earth relationship, like "wild," "pristine," "natural" and "ecological," inevitably incorporate some degree of human impact because of the long historical trajectory of human transformation of ecosystems and the dependency on initial conditions that constrain how systems evolve. This chapter provides an analysis and critique of the notion of a mutually enhancing human-Earth relationship and some of the related concepts and terms noted above, and I make proposals for how these concepts can provide a coherent normative foundation for ecological law.

What is a mutually enhancing human-Earth relationship?

Fostering a mutually enhancing human-Earth relationship is not yet a prevalent societal goal at any scale, as overarching commitments in major international texts to sustained economic growth make clear (United Nations 2012; United Nations 2015a; G20 2019). Instead, the Anthropocene highlights the extent to which the human enterprise has impaired and, with a dominant focus on economic growth, continues to impair the capacity of humanity and Earth's ecosystems to be mutually enhancing. Thus, the Anthropocene brings into sharp focus the need to move past the current dominant narrative of the human-Earth relationship and the human behaviors it promotes as good and worthwhile. As explained in Parts I and II, this dominant anthropocentric narrative gives primacy to economic growth and technological innovation, reinforced by legal systems that are built around strong notions of State sovereignty and private property rights (M'Gonigle and Takeda 2013; Burdon 2015; Capra and Mattei 2015). At the extreme, it is a narrative of humanity "'mastering' nature to reclaim Eden" (Merchant 2003: 3)—at least in Western worldviews that ground economics and other constructs in a paradigm, largely biblically based, of orderly, Cartesian human dominion over nature (Merchant 1980; Ost 2003; Ponting 2007; Capra and Mattei 2015). This narrative is fatally flawed because ultimately, the construction of artificial nature destroys the capacity of the ecosphere to support the flourishing of human and other life. In this narrative, the law of unintended consequences ultimately overwhelms the impossible quest to dominate, control and perfect nature thoroughly. The surprise and irreducible uncertainty inherent in the evolution of socio-ecological systems are too greatly ignored, and individual human preferences and technological optimism too greatly relied upon.

The Anthropocene also calls into question the purist ecocentric narrative that laments the "decline from a pristine earth to a paved, scorched, endangered world" (Merchant 2003: 2) and calls for the restoration of wild nature on Earth. The purist ecocentric narrative idealizes a pristine world without human impacts—and hence, without humans. In essence, it is a "cult of wilderness" (Miller 2001) narrative, fatally flawed for two main reasons. First, it ignores the significant extent to which humans have dominated Earth's ecosystems. More than three-quarters of Earth's ice-free land surface is made up of "anthropogenic biomes"—broad regions in which humans have altered the structure and function of ecosystems through, agriculture, forestry, urbanization and other interventions (Ellis and Ramankutty 2008; Alessa and Chapin 2008). Second, if applied from the local to the global level, at its logical extreme, the cult of wilderness narrative leaves no meaningful place for humans in Earth's life systems.

Both narratives collapse as well because they rely too much on "the linear approach of modern scientific thinking" (Merchant 2003: 4). A foundational narrative based on a mutually enhancing human-Earth relationship must be systems-based and situated between these two imperfect narratives of the ideal human-Earth relationship. Properly understood, human-inclusive ecocentric approaches must include a thriving human presence within a life-enhancing global ecosystem and must reject ecocentrism that is indifferent to human life. With this important caveat, systems-based, ecocentric approaches to law and other normative regimes will be needed to yield "[n]ew kinds of stories, new ways of thinking, and new ethics" (Merchant 2003: 4) for the Anthropocene, those that promote socio-ecological integrity and a mutually enhancing human-Earth relationship. Yet, the situation of these human-inclusive ecocentric approaches between the extremes of self-destructive human domination and humanity-free pristine nature presents its own challenges in defining ecological integrity, a mutually enhancing human-Earth relationship and related concepts.

Berry described a mutually enhancing human-Earth relationship as one that reflects that "in reality there is a single integral community of the Earth that includes all its component members whether human or other than human [each of which] has its own role to fulfill, its own dignity, its inner spontaneity" (Berry 1999: 4). It is a relationship in which "[e]very being enters into communion with other beings" (Berry 1999: 4). Berry aligns the transition to a mutually enhancing human-Earth relationship with the emergence of the Ecozoic Era, "the period when humans will be present to the planet as participating members of the comprehensive Earth community" (Berry 1999: 8). The mutually enhancing human-Earth relationship Berry envisioned would require a comprehensive transformation of social and political institutions, of educational systems and of the meaning of a host of economic terms, such as "progress" and "profit" (Berry 1999).

With regard to law, Berry put forth the overarching notion that "[e]cology is not part of the law; law is an extension of ecology" (Berry 1999: 84)—in other words, that law should reflect and maintain a human role within the broader community of life that is life-enhancing and respectful of the ecological role of other members of that community. He argued that "[t]o achieve a viable human-Earth situation a new jurisprudence must envisage its primary task as that of articulating the conditions for the integral functioning of the Earth process, with special reference to a mutually enhancing human-Earth relationship" (Berry 1999: 61). Under this new jurisprudence, the living and non-living elements of the Earth system would be treated as a commons "to be shared in proportion to need among all members of the Earth community" (Berry 1999: 61). Anthropocentric legal systems that give property rights and other rights exclusively to humans and juridical persons such as corporations would give way to a more ecocentric jurisprudence in which components of the Earth community all would have co-equal rights. Above all, every component of the Earth community would have a right to exist (Cullinan 2011) and a right "for habitat and the opportunity … to fulfill its role in the natural system to which it belongs" (Berry 1999: 80). In this new jurisprudence, all components of the Earth system, including humans, would have a mutual responsibility to respect the well-being of all the other components, and "[j]ustice would consist in carrying out this complex of creative relationships" (Berry 1999: 61–62).

Berry's approach implicitly endorses a dynamic, non-linear, systems view of the human-Earth relationship, which must be part of the foundation of a practicable narrative of long-term flourishing of humans and other living and non-living elements of Earth systems (Berry 1999; Merchant 2003; Capra and Mattei 2015). However, his proposal for a mutually enhancing human-Earth relationship lacks clear criteria for determining when the relationship is mutually enhancing, or for adjudicating conflicting rights claims among components of the Earth community. For example, without more clarity on the meaning of mutual enhancement, one could argue that as long as they are present on Earth, even in the current era of human degradation of many interacting processes within the Earth system (Ellis and Ramankutty 2008; Rockström et al. 2009; Steffen et al. 2015), humans are "participating members of the comprehensive Earth community" (Berry 1999: 8)—at times not very nice ones, but at times life-enhancing.

More refined notions of membership in the Earth community, such as notions of householding, membership and entropic thrift (Brown 2015), are needed to make this idea practicable. These notions reinforce the need for humans to see the Earth as a cherished home, to identify themselves as integral members in a community of life and to be cautious and careful in the use of material and energy needed to sustain them. As well, the proposal for rights of each member of the Earth community to fulfill its role in the functioning of natural systems invites a complex, multi-scale and cross-scale inquiry into what that role is, and what systems are "natural" at different scales. Studies of the extent of anthropogenic biomes in Earth's terrestrial systems call into question what "natural" means in this context (Ellis and Ramankutty 2008; Alessa and Chapin 2008).

Merchant (2003) frames the notion of a mutually enhancing human-Earth relationship in a way that avoids the flaws of pure trust in either: (1) human control and domination of nature, or (2) a vision of ideal nature that ultimately excludes humanity. She frames this middle path in terms of a partnership between humans and Earth, asserting "[a] partnership ethic holds that the greatest good for the human and nonhuman communities is in their mutual living interdependence" (Merchant 2003: 223). This partnership ethic rejects the hubris of the Enlightenment project of human mastery over nature. It is grounded in core precepts extending moral standing, equity and rights of inclusion and participation to non-human communities and species, as well as to all human communities and cultures (Merchant 2003). Although Merchant's partnership ethic leaves open questions as to the scope of notions such as equity and health, it broadly aligns with Berry's notion of a mutually enhancing human-Earth relationship. But, the conceptual core of hers and Berry's ethical vision necessarily implies that a notion such as ecosystem services (to humans) be complemented with the notion of human services to ecosystems. This notion of mutual service has virtually no firm toehold in contemporary law and policy. Yet, it must be part of a practicable approach to ecological law.

Several scholars have attempted to provide some of the missing detail in considering the implications of a mutually enhancing human-Earth relationship for law and legal systems, particularly through the concepts of "wild law" and "Earth Jurisprudence." The next section considers those concepts, including a more detailed discussion of the rights of nature.

Wild law, Earth jurisprudence and rights of nature

Inspired by Berry's cosmological orientation and his promotion of a mutually enhancing human-Earth relationship, Cullinan (2011) introduced the notion of "wild law:" jurisprudential law that is rooted in the scientific laws underlying the working of the universe (the "Great Jurisprudence"), that seeks to regulate human interaction with the Earth community and not the Earth community itself, and that gives Earth rights to both humans and other components and members of the Earth community to fulfill their role in the evolutionary processes of Earth systems. Consistent with Berry's orientation towards rights (Berry 1999), humans would be required to limit their rights so as to allow the Earth rights of all other components of the Earth community to be fully exercised.

The Great Jurisprudence and Earth rights are thus the central features of Cullinan's construction of wild law. To Cullinan, "Earth jurisprudence" (Cullinan 2011: 78, 170) is the expression of wild law in legal philosophies generated by humans that are consistent with the Great Jurisprudence. In this sense, the relationship of Earth jurisprudence to the Great Jurisprudence, cast as a kind of ultimate universal truth, is akin to the relationship of human laws to natural law in the natural law tradition. In the natural law tradition, "natural" law is tied to "divine law" and is overarching and beyond question (Aquinas 1988; Burdon 2015). Human law is always inferior to natural law because the human capacity to know divine truths is always imperfect—just as the irreducible uncertainty of the evolution of complex systems renders human law inferior to the Great Jurisprudence. Wild law and Earth jurisprudence may be seen as ecological transformations of Thomas of Aquinas's basic structure. Burdon (2015) is more specific than is Cullinan regarding the analogy of Earth jurisprudence to natural law theory, and more supportive. To Burdon, the overarching law is the Great Jurisprudence, or the great law (Berry 1999; Burdon 2015), and human law is the evolving best effort to apply the great law to specific situations. Earth jurisprudence to Burdon is the framework that maintains this relationship of the great law to human law and provides a moral basis for treating any human law inconsistent with the great law as "a corruption of law and not morally binding on a population" (Burdon 2015: 82).

In Cullinan's approach, Earth rights all derive from "the most fundamental source of all, the universe" (Cullinan 2011: 96). Building on Berry's description of the universe as "composed of subjects to be communed with, not as objects to be exploited" (Berry 1999: x-xi), in which humans have no special moral standing, Cullinan asserts that given their foundation in the order of the universe, Earth rights exist even if they are not legally enforceable. Thus, the general failure of a legal system to recognize rights of most non-human entities as legally enforceable "is simply an indication that the legal system is insufficiently developed to reflect the reality of their existence" (Cullinan 2011: 97). Because Earth rights of all components of the Earth community are reciprocal and co-equal in the way they function together as a whole, "the first principle of Earth jurisprudence must be to give precedence to the survival, health and prosperity of the whole Community over the interests of any individual or human society" (Cullinan 2011: 100). Thus, Cullinan asserts that human rights must be limited by the interests of the whole community, as well as by the rights of each other component of the Earth community to have "the liberty to fulfil its role within the Earth Community" (Cullinan 2011: 105)—also expressed as the right to fulfill its evolutionary role or to "exist in its natural state" (Cullinan 2011: 103).

Cullinan's conception of wild law, built on the foundation of the Great Jurisprudence and Earth rights, has strong appeal. He recognizes that to be consistent with an evolutionary, systems-based understanding of the world, wild law and Earth jurisprudence must be flexible and adaptive (Cullinan 2011: 26). However, wild law presents two main challenges. First, because concepts like evolution and the laws of the universe are value-neutral and not normative, additional criteria are needed to apply them to promote norms grounded in a mutually enhancing human-Earth relationship. Second, to avoid falling into the "cult of wilderness" trap (Cullinan 2011: 30), a practicable rights scheme that includes rights for all members of the Earth community must be clear as to how to resolve conflicts and draw lines between what is and what is not mutually enhancing in the human-Earth relationship.

As to the first challenge, wild law is ambiguous in regard to the normative significance it accords to the laws of the universe and evolutionary processes, as understood through contemporary science. Cullinan's critique of contemporary law seems to suggest that the human enterprise as currently oriented to give primacy to economic growth, individual preferences, commodification, monetization and technological innovation, is somehow defying laws of the universe or of evolution. However, evolution is inevitable; the laws of physics, ecology and other natural sciences are value-neutral, and evolution is an inevitable consequence of moving forward in time. Both a thriving tropical rainforest and an urban parking lot exist and evolve in accordance with the scientific laws of the universe. In their present state, which is the culmination of a complex history of social and ecological change at local to global scales, both represent an evolution consistent with scientific laws. Thus, the normative underpinning of wild law, which seeks enhancement of both humans and all other components of the Earth community, requires a more clear, detailed and nuanced description. Planetary boundaries and other metrics discussed in Chapter 10 provide plausible indicators of some key aspects of the mutually enhancing human-Earth relationship underlying wild law.

The relevant question is the extent to which the full range of potential evolutionary paths and future trajectories tend toward or are consistent with a mutually enhancing human-Earth relationship. This is in part a dialectical inquiry, in which moral or ethical imperatives about the human-Earth relationship are inevitably, but adaptively and iteratively, derived from scientifically informed social discourse as to what constitutes mutual human-Earth enhancement persistently across spatial and temporal scales. Because of these complex relationships of scale, this discourse inevitably occurs in an intricate web of a diverse set of cultural, political, economic and ecological contexts.

Because of the irreducible uncertainty and unpredictability of the evolution of complex social and ecological systems, empirical evidence can always emerge that calls into question moral precepts or that reveals irreconcilable conflicts among them (Elliott 2005). Working towards a mutually enhancing human-Earth relationship in practice requires the humility to acknowledge uncertainty, be vigilant about tracking relevant indicators and continually adapt policy to ensure respect for ecological limits as well as intragenerational, intergenerational and interspecies fairness and enhancement (Brown and Garver 2009).

The second challenge, again, is that Cullinan's proposed wild law sets out a scheme of rights for all living and non-living members of Earth's ecosystems without setting out clear criteria for determining how to resolve conflicts among them. Specifically, Cullinan's proposal does not clearly delineate the limits on human rights or describe criteria for determining when humans are fulfilling their "proper" evolutionary role. Both beavers and humans can alter watercourses so as to induce profound ecological change. Wild law seems to give beavers a pass, without clearly describing when humans have so altered a river as to breach the rights of the river and other components of the ecosystem of which it is a part.

Without clearer criteria, according rights to all components of the Earth community, or to "nature" generally, risks creating a system in which nothing and nobody has meaningful rights (Caldwell and Shrader-Frechette 1993). As well, the rights of nature may have different weight at different scales. For example, building a hydroelectric dam may appear to violate the rights of nature because of harm to local ecosystems, but to favor the rights of nature at the global scale because it would reduce greenhouse gas emissions (Rühs and Jones 2016). With such vagueness, conflicts might be decided by power and force rather than by law (Burdon 2015), or according to interpretations of law that end up eroding the long-term rights of nature in view of conflicting rights grounded in short-term social or economic interests.

For example, Burdon (2015) notes that Bolivia, Ecuador and over 30 jurisdictions in the United States have recognized rights of nature. However, these legislative gains have been achieved prior to a significant global shift in paradigm away from the growth-insistent narrative, in which profit-making ventures that rely on ecologically harmful practices generally have a particularly powerful influence in public decision-making. In such a context, a real danger exists that rights of nature will be interpreted in a way that does not ensure a mutually enhancing human-Earth relationship (Escobar 2010; Whittemore 2011).

The example of Ecuador is revealing. Ecuador adopted a new constitution in 2008 that says that "Nature, or Pacha Mama, where life is reproduced and occurs, has the right to integral respect for its existence and for the maintenance and regeneration of its life cycles, structure, functions and evolutionary processes"

(Constitution of the Republic of Ecuador, Title II, Chapter 7, Article 71 [English translation]).[1] Nonetheless, the Ecuadorian government subsequently opened the Yasuni National Park to oil exploitation after other countries failed to meet Ecuador's request for payments to compensate its proposed moratorium on such exploitation to reduce greenhouse gas emissions (Finer et al. 2010; Martin and Scholz 2014). This decision implied that Ecuador's constitutional rights of nature were contingent on other rights or considerations—in essence, that they were up for sale. Further, judicial interpretation of the rights of nature has already exposed divergent perspectives on the extent of nature's rights. In the first Ecuadorian court case dealing with the rights of nature, the Vilcabamba river case, the court held that the rights of nature meant that a precautionary approach was required when considering harms to nature (that is, a plaintiff could show potential harm and not only certain harm) and that the Party accused of breaching those rights has the burden of proof (Rühs and Jones 2016). However, it is not clear that the court interpreted the rights of nature to extend environmental protections much beyond those under Ecuador's existing environmental law (Rühs and Jones 2016). Furthermore, other Ecuadorian tribunals have adopted a narrower interpretation of the rights of nature (Rühs and Jones 2016).

Another concern about rights of nature is that they risk being trapped within an anthropocentric worldview because they emerged from rights-based legal frameworks developed within an anthropocentric worldview that perpetually impedes "sustainable solutions to our global crisis" (Capra and Mattei 2015: 76). For example, rights of nature are typically described as giving personhood to nature in the law (Boyd 2017: 36, 131). Undoubtedly, contemporary law is well-suited to framing rights of nature this way, and lawmakers and courts in several nations have already either legislatively adopted or adjudicated rights of nature within existing rights frameworks to which they were accustomed (Boyd 2017; Burdon and Williams 2016). Yet, giving nature personhood and rights essentially equips nature for battle with other rights holders and those who impact nature's rights in a way that—unlike nature—is divisive, reductionist and atomistic. To place holistic protection of nature firmly within a human-inclusive ecocentric worldview and better support the thriving and regeneration of Earth's whole community of life (Garver 2019), ecological law should be based more clearly and coherently on the integration of humans into nature. Instead of incorporating rights that give nature legal *personhood*, ecological law should ultimately be about adopting principles and rules that give people *naturehood* in the law (Braverman 2018: 135, citing Mari Margil).

Finally, experience with human rights to a healthy environment illustrates the risk of gradual erosion of rights of nature if they are not more firmly embedded in a human-inclusive ecocentric paradigm. A growing number of countries enshrine in their constitutions the right to a healthy or clean environment, which may imply the need to respect ecological limits (Hill et al. 2003: 381–389; Boyd 2017: xix). In Pennsylvania, the State supreme court has held that the State's constitutional right to a healthy environment (a more anthropocentric right regarding the human-Earth relationship), must be balanced against other constitutional and social and economic rights.[2] As a result, that State constitutional right has had little impact on decisions affecting the human-Earth relationship in Pennsylvania (Cussack 1993). This example, along with others (Cussack 1993), suggest that if rights of nature or similar rights are adopted within the dominant global paradigm of growth insistence, or without clear criteria for when ecological limits must be accorded primacy or how to interpret them in light of competing social objectives, the risk is high that the growth-insistent paradigm will overwhelm or at least dilute those rights (Whittemore 2011).

The concept of ecological integrity has been invoked to help give more clarity to broader concepts like sustainability or a mutually enhancing human-Earth relationship. Defining it, too, presents challenges, which are discussed in Chapter 12.

Notes

1 *Constitution of the Republic of Ecuador* (2008) arts. 71 and 72.
2 *Payne v. Kassab*, 312 A.2d 86, 94 (Pa. Commw. Ct. 1973), *aff'd,* 361 A.2d 263 (Pa. 1976).

References

Alessa, L., and S. Chapin III. 2008. Anthropogenic biomes: a key contribution to earth-system science. *Trends in Ecology and Evolution* 23(10): 529–531.

Alexander, S. 2011. Earth Jurisprudence and the Ecological Case for Degrowth, pp. 293–302 in P. Burdon, ed., *Exploring Wild Law: The Philosophy of Earth Jurisprudence.* Kent Town, South Australia: Wakefield Press.

Aquinas, Saint Thomas. 1988. *On Law, Morality, and Politics* (W.P. Baumgarth and R.J. Regan, eds.). Indianapolis IN: Hackett Publishing Co.

Berry, T. 1999. *The Great Work: Our Way Into the Future*. New York: Three Rivers Press.

Boyd, D.R. 2017. *The Rights of Nature: A Legal Revolution That Could Save the World.* Toronto: ECW Press.

Braverman, I. 2018. Law's underdog: a call for more-than-human legalities. *Annual Review of Law and Social Sciences* 14: 127–144.

Brown, P.G. 2015. Ethics for Economics in the Anthropocene, Chapter 2 in P.G. Brown and P. Timmerman, eds., *Ecological Economics for the Anthropocene*. New York: Columbia University Press.

Brown, P.G., and G. Garver. 2009. *Right Relationship: Building a Whole Earth Economy.* San Francisco: Berrett-Koehler.

Burdon, P.D., ed. 2011. *Exploring Wild Law: The Philosophy of Earth Jurisprudence.* Kent Town, South Australia: Wakefield Press.

Burdon, P.D. 2015. *Earth Jurisprudence: Private Property and the Environment.* New York: Routledge.

Burdon, P.D., and C. Williams. 2016. Rights of Nature: A Constructive Analysis, Chapter 8 in D. Fisher, ed., *Research Handbook on Fundamental Concepts of Environmental Law.* Cheltenham, UK: Edward Elgar Publishing.

Caldwell, L.K., and K.S. Shrader-Frechette. 1993. *Policy for Land: Law and Ethics.* Boston: Roman and Littlefield.

Capra, F., and U. Mattei. 2015. *The Ecology of Law: Toward a Legal System in Tune with Nature and Community.* San Francisco: Berrett-Koehler.

Cullinan, C. 2011. *Wild Law: A Manifesto for Earth Justice (2nd ed.).* White River Junction VT: Chelsea Green.

Cussack, M.E. 1993. Judicial interpretation of state constitutional rights to a healthful environment. *Boston College Environmental Affairs Law Review* 20(1): 172–201.

D'Alisa, G., F. Demaria and G. Kallis, eds. 2015. *Degrowth: A Vocabulary for a New Era.* New York: Routledge.

Daly, H.E. 2015. Economics for a Full World. *Great Transitions Initiative* [online]: www.tellus.org/pub/Daly-Economics-for-a-Full-World.pdf.

Elliott, H. 2005. *Ethics for a Finite World: An Essay Concerning a Sustainable Future.* Golden CO: Fulcrum Publishing.

Ellis, E.C., and N. Ramankutty. 2008. Putting people in the map: anthropogenic biomes of the world. *Frontiers in Ecology and the Environment* 6(8): 439–447.

Escobar, A. 2010. Latin America at a crossroads. *Cultural Studies* 24(1): 1–65.

Finer, M., R. Moncel and C.N. Jenkins. 2010. Leaving the oil under the Amazon: Ecuador's Yasuni-ITT initiative. *Biotropica* 42(1): 63–66.

G20. 2019. *G20 Osaka Leaders' Declaration* [online]: www.g20.utoronto.ca/2019/2019-g20-osaka-leaders-declaration.html.

Garver, G. 2019. Confronting remote ownership problems with ecological law. *Vermont Law Review* 43(3): 425–454.

Hill, B., S. Wolfson and N. Targ. 2003. Human rights and the environment: a synopsis and some predictions. *Georgetown International Environmental Law Review* 16: 359–402.

M'Gonigle, M., and L. Takeda. 2013. The liberal limits of environmental law: a green legal critique. *Pace Environmental Law Review* 30(3): 1005–1115.

Martin, P.L., and I. Scholz. 2014. *Ecuador's Yasuni-ITT Initiative: What Can We Learn from its Failure?* Articles and Debates 5.2. Geneva: The Graduate Institute.

Merchant, C. 1980. *The Death of Nature: Women, Ecology and the Scientific Revolution.* San Francisco: Harper & Row.

Merchant, C. 2003. *Reinventing Eden: The Fate of Nature in Western Culture.* New York: Routledge.

Miller, C. 2001. An open field. *Pacific Historical Review* 70(1): 69–76.

Ost, F. 2003. *La nature hors la loi: L'écologie à l'épreuve du droit.* Paris: La Découverte/Poche.

Ponting, C. 2007. *A New Green History of the World: The Environment and the Collapse of Great Civilizations.* Toronto: Penguin Books.

Raworth, K. 2017. *Doughnut Economics: Seven Ways to Think Like a 21st Century Economist.* White River Junction, VT: Chelsea Green Publishing.

Rockström, J., W. Steffen, K. Noone, Å. Persson, F.S. Chapin III, E. Lambin, T.M. Lenton, M. Scheffer, C. Folke, H. Schellnhuber, B. Nykvist, C.A. De Wit, T. Hughes, S. van der Leeuw, H. Rodhe, S. Sörlin, P.K. Snyder, R. Costanza, U. Svedin, M. Falkenmark, L. Karlberg, R.W. Corell, V.J. Fabry, J. Hansen, B. Walker, D. Liverman, K. Richardson, P. Crutzen and J. Foley. 2009. Planetary boundaries: exploring the safe operating space for humanity. *Ecology and Society* 14(2): 32 [online] URL: www.ecologyandsociety.org/vol.14/iss2/art32/

Rühs, N., and A. Jones. 2016. The implementation of Earth jurisprudence through substantive constitutional rights of nature. *Sustainability* 8: 174.

Sieferle, R.P. 2001. *The Subterranean Forest: Energy Systems and the Industrial Revolution.* Cambridge UK: The White Horse Press.

Steffen, W., K. Richardson, J. Rockström, S.E. Cornell, I. Fetzer, E.M. Bennett, R. Biggs, S.R. Carpenter, W. de Vries, C.A. de Wit, C. Folke, D. Gerten, J. Heinke, G.M. Mace, L.M. Persson, V. Ramanathan, B. Reyers and S. Sörlin. 2015. Planetary boundaries: guiding human development on a changing planet. *Science* 347(6223): 1259885-1–1259885-10.

United Nations. 2012. *The future we want.* New York: United Nations [online]: https://documents-dds-ny.un.org/doc/UNDOC/GEN/N11/476/10/PDF/N1147610.pdf?OpenElement.

United Nations. 2015a. *Transforming Our World: The 2030 Agenda for Sustainable Development.* New York: United Nations.

Whittemore, M.E. 2011. The problem of enforcing nature's rights under Ecuador's constitution: why the 2008 environmental amendments have no bite. *Pacific Rim Law & Policy Journal* 20(3): 659–691.

9 The systems-based perspective underlying ecological law

The Anthropocene resonates forcefully at the many junctions and points of integration of the humanities and social sciences with the natural and physical sciences, all of which are subject to the complex evolution of dynamic, interactive ecological and socially constructed systems. As well, the Earth system and its subsystems are all infused with the mysteries of irreducible uncertainty and unpredictability, emergence and the potential for catastrophic shifts. Legal systems are no exception. However, as with other socially constructed systems, the variable of human intention plays a leading role in how law has affected and will affect the human prospect. In this chapter, I analyze how a systems-based perspective of law as a complex adaptive system that interacts and evolves along with other systems in a global panarchy (Gunderson and Holling 2002; Ruhl 2012) can provide the foundation for a transition to a mutually enhancing human-Earth relationship in the Anthropocene.

Legal systems act as complex adaptive systems that interact with other natural and socially constructed complex adaptive systems in the global panarchy across multiple temporal and spatial scales (Ruhl 2012). Through a long historical evolution of these interactions, contemporary legal systems developed largely in support of the worldviews and politics that have become more and more hardened at all scales around insistence on economic growth, the paramount power of sovereign states and strong private property rights, all undergirded by a primary reliance on utilitarianism, reductionist thinking and an assumption of linearity in cause-and-effect interactions (McLellan et al. 2014; Capra and Mattei 2015). Understanding how legal systems behave as complex adaptive systems and interact with other such systems may yield an adaptive approach to law that is more responsive to and coherent with processes of global environmental change that are factoring more and more heavily in the human prospect. Such an understanding should enhance the potential to transition toward law that promotes a mutually enhancing human-Earth relationship.

Legal systems as complex adaptive systems

Panarchy theory aims to describe how complex adaptive systems interact and evolve at multiple temporal and spatial scales, through repeating cycles made up of phases of exploitation, conservation, release and reorganization (Holling et al. 2002; Westley et al. 2013). It can be applied in attempting "to improve the design of policy responses [to declines in environmental quality and the growing scale of human activities]" (Ruhl 2012: 1). Law is a necessary element in applying panarchy theory so as to implement desired policy responses (Ruhl 2012). Several scholars have been building the case that legal systems themselves are complex adaptive systems that must be accounted for in panarchy theory and social-ecological systems theory (Ruhl 1996; Hornstein 2005; Ruhl 2008; Ruhl 2012; Craig 2013; Garmestani and Allen 2014). Because "law is where humans write the rules for other social systems" (Ruhl 2008: 897), applying panarchy theory to address the current human-driven global ecological crisis may well "require managing the complex adaptive legal system to adaptively manage other complex adaptive natural and social systems, all in a way that maintains some level of social order" (Ruhl 2012: 2).

An adaptive system is one in which the agents or components of the system make adjustments based on what occurs in their interactions. Complexity occurs in an adaptive system when the agents and elements develop non-linear interdependencies, the dynamics of which evolve as the agents and elements adapt according to their interactions and external inputs, giving the system as a whole characteristics that transcend the characteristics of its individual parts (Hornstein 2005; Ruhl 2008; Craig 2013). Complex adaptive systems are emergent, which means that although their components interact according to deterministic rules, the system-wide results of these numerous interactions are continually novel, non-linear, not entirely predictable and characteristic of the system as a whole rather than of its individual components (Hornstein 2005; Ruhl 2008). Emergence explains in part why attempts to control complex adaptive systems often, or perhaps inevitably, lead to unintended consequences. Complex adaptive systems are also path dependent, such that the past trajectory or the initial conditions of the system constrain the range of its possible futures (Hornstein 2005; Ruhl 2008). Emergence, non-linearity and path dependence have been identified as particularly important for considering how legal systems evolve and adapt along with the systems with which they interact (Hornstein 2005).

Incorporating human understanding, intention and design complicates the extension of complexity theory from natural ecosystems to social constructs such as legal systems (Ruhl 1996; Ruhl 2008; Cudworth and Hobden 2013). Human decision-making, both individually and collectively, can itself be considered a complex adaptive system that integrates evolving epistemological and ontological foundations and the knowledge and understanding built upon them (Cudworth and Hobden 2013). The interaction of human systems of knowledge, understanding, intention and decision-making with ecological systems at various scales can lead either to local or civilizational collapse, at one extreme, or to long-term resilience and adaptiveness, at the other (Ostrom 1990; Tainter 1990; Diamond 2005; Beddoe et al. 2009; Cudworth and Hobden 2013). The Anthropocene concept captures the long-term and global scale of the capacity of human decisions to affect Earth's ecological conditions, as well as the looming potential for systems of human thought to lead either to civilizational or societal collapse on a global scale or to the long-term resilience and adaptiveness inherent in a mutually enhancing human-Earth relationship.

From the Code of Hammurabi and other early formal legal systems to contemporary legal systems, law is a key reflection of the ordering of human knowledge, understanding and decision-making in support of societal goals at different scales. Human intention and agency distinguish legal and other socially constructed systems from ecosystems (Cudworth and Hobden 2013). Legal scholars have developed a plausible case that legal systems nonetheless behave like complex adaptive systems, oriented around a core tension among interdependent and enforceable rights, freedoms, obligations and prohibitions (Ruhl 1996).[1] The balance among these variables depends on, and in turn influences, the evolving interaction of legal systems with other systems, in particular economic, political, cultural, social, ecological and technological systems (M'Gonigle and Takeda 2013). Green Legal Theory (M'Gonigle 2008) implicitly adopts this systems perspective in taking a pluralistic view of law in which cultural context that transcends formal law expands the systemic understanding of and reliance on rules of conduct. Table 9.1 summarizes key elements of the multi-dimensional, multi-scalar and interdependent context and structure of legal systems.

Ruhl (2008) provides detailed examples of how legal systems exhibit both the agent properties (heterogeneity, deterministic rules, non-linearity, feedback mechanisms) and system properties (emergence, path dependence, self-organization, critical states, adaptive resistance and resilience, phase transitions) of complex adaptive systems. For example, the actors and institutions involved in legal systems—lawyers, clients, citizens, voters, elected officials, legislators, enforcers, judges, courts, legislatures, administrative bodies and so on—are diverse and heterogeneous; judicial decisions and new statutes or constitutions can bring about sudden, major—that is, non-linear—shifts in the legal landscape; the full meaning of a particular law or legal rule emerges as the legal system evolves, through the not entirely predictable interplay of the actors and processes involved; and the existence of a particular constitution or set of laws, as well as the context in which laws are developed and implemented, has a strong effect on the possible future states of the legal system (Ruhl 2008).

Much of the scholarship on legal systems as complex adaptive systems has arisen in the environmental arena, with a focus on how environmental law might (or should) evolve to become more coherent with a systems-based understanding of how law interacts with other systems. Yet, focusing only on environmental law in the analysis of law as a complex adaptive system that interacts with other systems in a global panarchy is inherently inconsistent with a systems approach (M'Gonigle 2008). A fully integrated and holistic, systems-based treatment of law and its relation to the global panarchy of complex adaptive systems requires a deep and comprehensive analysis of every corner of the law—including all the arenas in Table 9.1.

Table 9.1 Structure and key elements of legal systems

System feature	*Key elements*
Context within the global panarchy	• Worldview • Relationship with economic, political and socio-cultural systems • Relationship with ecological systems • Relationship with technological systems
Legal mechanisms	• Constitutions • Codes and statutes • Bills of rights • Executive decrees, orders, edits, etc. • Administrative rules, regulations and orders • Judicial decisions (common law, civil law, etc.) • Treaties and international agreements • Other international law • Legal plurality: culture, tradition, custom, etc. that informally yet consistently governs behavior (M'Gonigle 2008)
Actors, agents and power structures (may exist and interact at all scales from global to local)	• Governance regime (democracy, dictatorship, etc.) • Executive/administrative (heads of State, ministers, law enforcers, regional/local executives, etc.) • Legislators/parliamentarians • Judicial/quasi-judicial (judges, arbitrators, etc.) • Commissions, independent bodies, auditors, etc. • Individual citizens • Juridical "persons" with rights (e.g. corporations, "nature") • Civil society • Public/governmental entities
Key arenas (may exist and interact at all scales, from global to local)	• Basic rights and responsibilities • Property (private, public, common, things not "owned") • Finance, tax and money systems • Contracts • Criminal justice and tort liability (non-criminal) • Corporations • Commerce and international trade • Environment and biodiversity • Energy and natural resources • Health and social safety net • Education • Labor • Agriculture and food systems • Cities and urbanization • Transportation • International relations and security (e.g. armed conflict) • Art and creativity

A consensus is apparent in the literature analyzing law as a complex adaptive system that, to be more coherent with a systems-based approach and to regulate human impacts on the ecosystems on which humanity depends, legal systems must become more explicitly flexible and adaptive (Ruhl 1996; Craig 2013; Garmestani and Allen 2014). To function properly, this approach requires a coherent and adequately supported program of monitoring and response, taking a precautious approach (Garver 2013). Indeed, adaptive management and precaution against undermining ecological integrity are key features of such a legal system (Hornstein 2005; Ruhl 2012). Adaptive management approaches give much greater emphasis to back-end adaptation to the results of monitoring, as opposed to front-end process and impact analysis, than is common in contemporary environmental law (Ruhl 2012; Garver 2013; Garmestani and Allen 2014). Precaution against undermining ecological integrity or crossing critical ecological thresholds such as those associated with planetary boundaries (Rockström et al. 2009; Steffen et al. 2015) is appropriate in light of the irreducible uncertainty and inevitable potential for unintended consequences that exist in complex adaptive systems. A precautionary approach implies a need to focus adaptive management on the management of humans and their social constructs and physical undertakings, rather than on the structure and function of the ecosystems in which they are embedded—although some direct ecosystem management is inevitable to remedy human disruptions of ecosystems and to promote ecological restoration and regeneration. The application of adaptive and precautionary approaches is analyzed further in Chapter 11.

Note

1 Ruhl (1996) does not include obligations, yet they round out the full range of interrelated legal provisions that define what a person (1) can insist on doing or being protected against (rights), (2) may do (freedoms), (3) must do (obligations), or (4) may not do (constraints or prohibitions).

References

Beddoe, R., R. Costanza, J. Farley, E. Garza, J. Kent, I. Kubiszewski, L. Martinez, T. McCowen, K. Murphy, N. Myers, Z. Ogden, K. Stapleton and J. Woodward. 2009. Overcoming systemic roadblocks to sustainability: the evolutionary redesign of worldviews, institutions, and technologies. *Proceedings of the National Academy of Science* 106(8): 2483–2489.

Capra, F., and U. Mattei. 2015. *The Ecology of Law: Toward a Legal System in Tune with Nature and Community*. San Francisco: Berrett-Koehler.

Craig, R.K. 2013. Learning to think about complex environmental systems in environmental and natural resources law and legal scholarship: a twenty-year retrospective. *Fordham Environmental Law Review* 24: 87–102.

Cudworth, E., and S. Hobden. 2013. Complexity, ecologism, and posthuman politics. *Review of International Studies* 39: 643–664.

Diamond, J. 2005. *Collapse: How Societies Choose to Fail or Succeed*. New York: Viking Press.

Garmestani, A.S., and C.R. Allen, eds. 2014. *Social-Ecological Resilience and the Law*. New York: Columbia University Press.

Garver, G. 2013. The rule of ecological law: the legal complement to degrowth economics. *Sustainability* 5(1): 316–337 [online]: http://mdpi.com/2071-1050/5/1/316/htm

Gunderson, L.H. and C.S. Holling, eds. 2002. *Panarchy: Understanding Transformations in Human and Natural Systems*. Washington, DC: Island Press.

Holling, C.S., L.H. Gunderson and D. Ludwig. 2002. In Quest of a Theory of Adaptive Change, pp. 3–22 in L.H. Gunderson and C.S. Holling, eds. *Panarchy: Understanding Transformations in Human and Natural Systems*. Washington, DC: Island Press.

Hornstein, D.T. 2005. Complexity theory, adaptation, and administrative law. *Duke Law Journal* 54: 913–960.

M'Gonigle, M. 2008. Green legal theory: a new approach to the concept of environmental law. *Neue Konzepte* 4: 34–38.

M'Gonigle, M., and L. Takeda. 2013. The liberal limits of environmental law: a green legal critique. *Pace Environmental Law Review* 30(3): 1005–1115.

McLellan, R., L. Iyengar, B. Jeffries and N. Oerlemans, eds. 2014. *Living Planet Report 2014: Species and Spaces, People and Places*. Gland, Switzerland: WWF.

Ostrom, E. 1990. *Governing the Commons: The Evolution of Institutions for Collective Action*. New York: Cambridge University Press.

Rockström, J., W. Steffen, K. Noone, Å. Persson, F.S. Chapin III, E. Lambin, T.M. Lenton, M. Scheffer, C. Folke, H. Schellnhuber, B. Nykvist, C.A. De Wit, T. Hughes, S. van der Leeuw, H. Rodhe, S. Sörlin, P.K. Snyder, R. Costanza, U. Svedin, M. Falkenmark, L. Karlberg, R.W. Corell, V.J. Fabry, J. Hansen, B. Walker, D. Liverman, K. Richardson, P. Crutzen and J. Foley. 2009. Planetary boundaries: exploring the safe operating space for humanity. *Ecology and Society* 14(2): 32 [online] URL: www.ecologyandsociety.org/vol.14/iss2/art32/

Ruhl, J.B. 1996. Complexity theory as a paradigm for the dynamical law-and-society system: a wake-up call for legal reductionism and the modern administrative state. *Duke Law Journal* 45(8): 849–928.

Ruhl, J.B. 2008. Law's complexity: a primer. *Georgia State University Law Review* 24(4): 885–911.

Ruhl, J.B. 2012. Panarchy and the law. *Ecology and Society* 17(3); art 31 [online]: http://dx.doi.org/10.5751/ES-05109-170331.

Steffen, W., K. Richardson, J. Rockström, S.E. Cornell, I. Fetzer, E.M. Bennett, R. Biggs, S.R. Carpenter, W. de Vries, C.A. de Wit, C. Folke, D. Gerten, J. Heinke, G.M. Mace, L.M. Persson, V. Ramanathan, B. Reyers and S. Sörlin. 2015. Planetary boundaries: guiding human development on a changing planet. *Science* 347(6223): 1259885-1–1259885-10.

Tainter, J.A. 1990. *The Collapse of Complex Societies*. Cambridge UK: Cambridge University Press.

Westley, F.R., O. Tjornbo, L. Schultz, P. Olsson, C. Folke, B. Crona and O. Bodin. 2013. A theory of transformative agency in linked social-ecological systems. *Ecology & Society* 18(3): 27.

10 The limits-insistent narrative and planetary boundaries

Ecological law is emerging from the dynamic tension between the dominant growth-insistent narrative and a limits-insistent narrative that aligns with a mutually enhancing human-Earth relationship. The limits-insistent narrative is central to the Club of Rome's notion of limits to growth (Meadows et al. 2004), as well as disciplines like ecological economics (Boulding 1975; Daly 1996) and social movements like the degrowth movement (Kallis 2018). Living within limits is also a core concept in ecological law, and it means that the goal of enclosing the social and economic spheres within the Earth's ecological limits must have primacy against other goals (Garver 2013). Planetary boundaries of safe operating space for humanity (Rockström et al. 2009) are a key building block for moving the limits-insistent narrative from theory, where it largely remains, to practice.

An international team of researchers led by Johan Rockström of the Stockholm Resilience Centre first proposed planetary boundaries of safe operating space for humanity in 2009, and they suggested planetary boundaries might serve as the basis for "novel and adaptive governance approaches at global, regional and local scales" (Rockström et al. 2009: 6). The boundaries are systems-based global ecological limits beyond which humanity faces the loss of safe operating space—that is, an unacceptable "risk of deleterious or even catastrophic environmental change at continental to global scales" (Rockström et al. 2009: 28). In 2015, planetary boundaries researchers updated the boundaries, taking into account the explosion of comments and critiques about the nine boundaries initially proposed (Steffen et al. 2015). The updated boundaries are, at the global scale[1] (Steffen et al. 2015):

- **Climate change**: limit carbon dioxide in the atmosphere to 350 ppm (zone of uncertainty: 350–450 ppm) and net radiative forcing to +1 watt per meter squared (zone of uncertainty: +1 to +1.5 watt per meter squared);
- **Change in biosphere integrity**: limit species extinctions to no more than ten extinctions per million species per year (E/MSY) (zone of uncertainty: 10 to 100 E/MSY);
- **Stratospheric ozone depletion**: limit the loss to no more than 5 percent of pre-industrial levels (zone of uncertainty: 5 to 10 percent);

- **Ocean acidification**: maintain at least 80 percent of the pre-industrial level of aragonite in the surface waters of the oceans (zone of uncertainty: 70 to 80 percent);
- **Biogeochemical flows:** limit the flow of phosphorus to the oceans to no more than eleven teragrams of phosphorus per year (TgP/year) (zone of uncertainty: 11 to 100 TgP/year), and limit industrial and intentional biological fixation of nitrogen to 62 teragrams per year (TgN/year) (zone of uncertainty: 62 to 82 TgN/year);
- **Land-system change**: maintain total forested land to at least 75 percent of the original forest cover (zone of uncertainty: 54 to 75 percent), based on maintaining at least 85 percent of original tropical and boreal forest cover and 50 percent of original temperate forest cover;
- **Freshwater use**: limit consumption of freshwater (blue water) to no more than 4,000 cubic kilometers per year (zone of uncertainty: 4,000 to 6,000 cubic km per year).

For atmospheric aerosol loading and introduction of novel entities, the researchers did not propose a boundary in either 2009 or 2015 (Rockström et al. 2009; Steffen et al. 2015).

As conceived, the planetary boundaries are "non-negotiable" (Rockström et al. 2009: 4). Rockström et al. (2009: 7) emphasize that "[t]he thresholds in key Earth System processes exist irrespective of peoples' preferences, values or compromises based on political and socio-economic feasibility, such as expectations of technological breakthroughs and fluctuations in economic growth." Thus, arguing that establishing a legal and policy regime that strictly respects the planetary boundaries is not economically or politically feasible is tantamount to arguing that ensuring the prospect for humanity's long-term safe operation is not economically or politically feasible—in essence, that ensuring economic and political feasibility is not economically and politically feasible! Ecological law therefore calls for establishing the planetary boundaries and then for determining how to adjust economic and political policy so as to respect the boundaries.

As noted earlier, global ecological limits imply a need to give primary normative importance to the acceptable risk of catastrophic ecological consequences for human and other life on the planet, as compared to risks based on social, political or economic concerns. Although planetary boundaries or limits can be expressed at a specific level, such as 350 ppm carbon dioxide in the atmosphere, they represent more fluid systems thresholds based on shifts in global system dynamics or on the aggregation of dispersed regional ecological impacts of human activity. These systems thresholds are impossible to pinpoint with certainty because of the non-linear dynamics of the human-Earth system, in which change can occur chaotically, in lurches, and the system can transcend points of no return long before feedbacks take effect so as to make the consequent impacts manifest. Adopting a precautionary approach, planetary boundary researchers have designated zones of certainty in which critical global systems thresholds are most likely to exist for each boundary category, with the safe boundary set at the most cautious extreme of the zone of uncertainty (Rockström et al. 2009; Steffen et al. 2015).

To be clear, normative boundaries, which are linked to but distinct from systems thresholds, are a human construct with implicit judgments as to the points beyond which the risk of exceeding systems thresholds or aggregate pressures is unacceptable (Rockström et al. 2009: 5). Thus, boundaries establish a bridge between the vast complexities of the underlying science and the normative arenas in which people operate individually and collectively and have to make decisions. For example, the proposed safe planetary boundary for atmospheric levels of carbon dioxide is set with a margin of safety to avoid a point-of-no-return threshold that is based on solid scientific understanding of the myriad interrelated consequences of increasing global average temperatures, such as increases in sea level and extensive loss of biodiversity (Hansen et al. 2008). The threshold cannot be determined with absolute certainty, but the boundary reflects a choice to operate safely below the threshold. Of course, the concept of a limit or boundary might imply that all is well until the limit or boundary is reached, whereas significant and undesirable deterioration in ecological or human well-being might occur well below the limit or boundary. Nonetheless, the planetary boundaries are grounded on recognition of the essential reliance of humanity on Earth's life systems, including other forms of life, and they incorporate a precautionary approach that should lessen concerns about living too close to the edge.

A challenge in designating global ecological limits is to give meaning to sub-global distribution of the drivers of ecological change and of the way that change is manifested. For example, human-induced loss of biodiversity and anthropogenic loadings of pollutants vary considerably from one locality or region to another, and the point at which they constitute in the aggregate a catastrophic and globally significant disruption of ecological systems is elusive. Indeed, the viability of some planetary boundaries as initially proposed, such as those related to nutrient loading and loss of biodiversity, was questioned because they are difficult to conceptualize or model at the global level, as opposed to regionally or locally (Nordhaus et al. 2012; Nyquist et al. 2013). Planetary boundary researchers have since developed a two-tiered approach for boundaries for anthropogenic nitrogen and phosphorous loading, land-system change, freshwater use and atmospheric aerosol loading to account for regional heterogeneity (Steffen et al. 2015). More specifically,

> The sub-global-level units of analysis for these six boundaries are not identical; they vary according to the role that the processes play in the Earth System: (i) changes in biosphere integrity occur at the level of land-based biomes, large freshwater ecosystems or major marine ecosystems as the largest sub-global unit; (ii) the role of direct, human-driven land-system change in biophysical climate regulation is primarily related to changes in forest biomes; (iii) freshwater flows and use occur at the largest sub-global level in the major river basins around the world; and (iv) changes in biogeochemical flows, exemplified by phosphorus and nitrogen cycling aggregate

> from relatively localized but very severe perturbations in intensive agricultural zones to affect global flows of nutrients.
>
> (Steffen et al. 2015: 3)

Other boundaries, in particular the planetary boundaries for climate change, stratospheric ozone depletion, ocean acidification and biosphere integrity are clearly of global significance. Thus, planetary boundary researchers have determined that climate change and biosphere integrity (expressed by two indicators, species extinction rate and the Biodiversity Intactness Index) are hierarchically "core boundaries" because

> they are highly integrated, emergent system-level phenomena that are connected to all of the other PBs[; t]hey operate at the level of the whole Earth System, and have co-evolved for nearly 4 billion years[; and t]hey are regulated by the other boundaries and, on the other hand, provide the planetary-level overarching systems within which the other boundary processes operate.
>
> (Steffen et al. 2015: 6–7)

By contrast, the other boundary processes operate through these two core boundaries, such that

> [t]he crossing of one or more of all of the other boundaries may seriously affect human wellbeing, and may predispose the transgression of a core boundary(ies), but does not by itself lead to a new state of the Earth System.
>
> (Steffen et al. 2015: 7)

Where, as with the planetary boundaries, several boundaries are used to express the aggregate human ecological impact on Earth, the boundaries' interrelatedness and interactions must be considered. Noting that "[i]nteractions among planetary boundaries may shift the safe level of one or several boundaries" (Rockström et al. 2009: 24), the initial planetary boundaries research team set each of their proposed boundaries on the assumption that no other boundary was transgressed. If this assumption does not hold, they suggest that the boundaries would most likely have to be adjusted so as to shrink humanity's safe operating space (Rockström et al. 2009). Thus, the interactions among different planetary boundaries have important implications. For example, concerns regarding land use are ultimately founded on the effects of land use on other boundaries, in particular biodiversity loss, climate change and nitrogen loading (Nordhaus et al. 2012; Steffen et al. 2015). As well, ocean acidification is directly related to increased levels of atmospheric carbon dioxide, for which the overarching consequence of concern is climate change. These relationships have an effect on the development of indicators, for example in determining relative priorities among impacts in regard to their contributions to the most urgent pressures on the global ecosystem. Considerations such as these regarding the relationship of boundaries to each other led planetary boundary researchers to designate climate change and biosphere integrity as core indicators, as explained above (Steffen et al. 2015).

The dynamics and interregional interdependencies of biogeochemical, hydrologic, climatic, atmospheric and other processes also prevent any sub-global region or locality from being isolated from the rest of the integrated global ecosystem, as a previously mentioned study that detected specific dioxins from sources in Mexico, the United States and southern Canada in mothers' milk and other receptors in Inuit communities in northern Canada illustrated starkly (Commoner et al. 2000). Moreover, humans or their impacts cover the entire globe. Human travel and migrations, along with the human-induced movement of other species, material and energy in an increasingly globalized economy, makes it unlikely that any region could achieve sustainability or other societal goals independent of the rest of the world. Agricultural land use, and other resource use, increasingly involves "interregional problem shifting" (Bringezu et al. 2012) as trade and human activities globalize.

The global and regional processes underlying the planetary boundaries, and their interaction, underscore these interdependencies (Rockström et al. 2009). All the planetary boundaries, even those involving global-scale processes like climate change and stratospheric ozone depletion, exhibit geospatial variability as to their impacts and socio-economic drivers. Some boundary categories, in particular land use change, freshwater use, biodiversity loss and nutrient loading, primarily involve regional or local processes and are included as global boundaries because either the aggregate scale of the sub-global impacts can affect global processes with critical systemic thresholds, like climate change, or because the undesirability of negative regional and local impacts may become so widespread as to have global importance (Rockström et al. 2009; Steffen et al. 2015). Further, interactions among boundaries are important at both global and sub-global scales. For example, deforestation in the Amazon—an important driver of biodiversity loss and climate change—can also affect water availability in Asia because of the interregional importance of tropical forests to global hydrological cycles and energy balances (Rockström et al. 2009).

Planetary boundaries and considerations of equity and fairness

The planetary boundaries are keyed to human safety and survival and do not provide much information about distribution and efficiency within those ecological limits, and therefore they are not as clearly committed to interspecies equity between humans and other living beings as they might be (Häyhä et al. 2016). Indeed, planetary boundary researchers acknowledge that the boundaries framework "does not as yet account for regional distribution of the impact, nor of its historical patterns [and does not] take into account the deeper issues of equity and causation" (Steffen et al. 2015: 7). Although safe operating space as developed by planetary boundaries researchers is consistent with a broad, human-inclusive ecocentric approach in that it aims to ensure "the flexibility to choose a myriad of pathways for human well-being and development" (Rockström et al. 2009), more detail regarding ways to ensure well-being for human and non-human life is warranted.

Ensuring well-being in an ecologically finite world means providing all present and future members of life's commonwealth "bounded capabilities" to flourish, contingent on fair sharing of Earth's limited capacity to support life (Jackson 2009: 45–47; Brown and Garver 2009). The notion of *bounded* capabilities adds nuance to Amartya Sen's notion of human capabilities to be well nourished, to live long lives and to engage meaningfully in society (Sen 2005), by underscoring the need to condition capabilities and freedom on aggregate ecological limits (Jackson 2009). Raworth (2017) builds out this idea by combining the nine planetary boundaries with 12 basic human needs for social justice and well-being in her proposal for "doughnut economics." The 12 categories of basic human needs are water, food, health, education, income and work, peace and justice, political voice, social equity, gender equality, housing, networks and energy (Raworth 2017: 38). "Below the social foundation lie shortfalls in human well-being, faced by those who lack life's essentials such as food, education and housing" (Raworth 2017: 39). This is the hole in the doughnut. "Beyond the ecological ceiling lies an overshoot of pressure on Earth's life-giving systems, such as through climate change, ocean acidification and chemical pollution" (Raworth 2017: 39). Those are the planetary boundaries. "But between these two sets of boundaries lies a sweet spot—shaped unmistakably like a doughnut—that is both an ecologically safe and socially just space for humanity" (Raworth 2017: 39).

For distribution, efficiency and other matters relevant to social constructs such as law, broader notions such as a mutually enhancing human-Earth relationship provide criteria that overlap with but go beyond those related to scale and planetary boundaries. Nonetheless, ensuring a mutually enhancing human-Earth relationship, as well as similar overarching orientations such as right relationship (Brown and Garver 2009) or strong sustainability (Halsnæs et al. 2007; Bosselmann 2008), is contingent at a minimum on ensuring safe operating space for humanity, the core objective behind the planetary boundaries concept (Rockström et al. 2009; Steffen et al. 2015). Moreover, because resilience is a key criterion of safe operating space (Rockström et al. 2009), the planetary boundaries approach also has commonalities with right relationship and with the notion that "[a] resilient economy—capable of resisting external shocks, maintaining people's livelihoods and living within our ecological means—is the goal we should be aiming for" (Jackson 2009: 157). Here, the main focus is on systems-based ecological metrics and indicators of the aggregate scale of human activity, with some but less attention to distribution and efficiency.

Metrics of performance in achieving societal goals, such as a mutually enhancing human-Earth relationship, would highlight socio-ecological patterns in some regions that could provide models for other regions, just as environmental history might "point the way to a better future by revealing the existence of societies in the past that had managed their relations with the environment more successfully, perhaps even sustainably" (McNeill 2003: 15). For example,

Pollard et al. (2010) illustrate the correlation of two metrics that are relevant to sustainability or other societal goals: human ecological footprint and the Human Development Index. Although both indicators have important shortcomings, this correlation provides some indication of countries in which, on average, social and economic development is being achieved with relatively low aggregate demand on the Earth's biocapacity (primarily in Latin America), and which may therefore provide useful models of the human-Earth relationship for less successful countries, such as the United States and Canada.[2] The correlation is consistent with a study of the performance of individual nations with respect to the planetary boundaries, which made initial and somewhat cautious findings that

> less developed countries now perform well, and in per capita terms use sustainable amounts of resources with respect to the [planetary] boundaries. In contrast, highly developed countries and some emerging economies transgress several of their national boundaries, although there is a less clear pattern for some others (e.g., biodiversity loss). In general, the performance of highly developed countries including Sweden is worse if consumption "footprints" rather than strictly territorial emissions/resource use are considered. The clear pattern associated with level of income for many of the boundaries cannot be ignored in the light of calls for the "right to develop" within the shared environmental space, and suggests that consumption patterns in highly developed countries need to be dealt with.
>
> (Nyquist et al. 2013: 11)

The study concludes that with respect to performance based on the planetary boundaries,

> it is most important to work with developed countries and countries with rapidly growing economies [because] these countries have higher absolute and per capita impacts on the environment globally, and thus a bigger responsibility for progressive action on, e.g., mitigating climate change.
>
> (Nyquist et al. 2013: 13)

Use of planetary boundaries and other global ecological limits for policy, law and governance

A systems indicator is only meaningful and useful if it conveys information relevant to a stated objective for the system. Thus, if the paramount objective of the global economy were to ensure continuous growth, then the combined Gross Domestic Product (GDP) of the world's nations presumably would present meaningful and useful information. By contrast, Brown and Garver (2009) contemplate an economy of which the purpose is to preserve and enhance the integrity, resilience and beauty of the commonwealth of life. A set of indicators of the extent to which this objective is being attained, and with potential use in supporting policies for attaining it, would be quite different from GDP. Similarly, indicators at the local to the global scale that support law, policy and governance that will lead to and maintain a mutually enhancing human-Earth relationship will inevitably be different from those, like GDP, that support dominant regimes of law, policy and governance currently.

Indicators of global ecological limits, such as planetary boundaries, can help frame the transition toward mutual human-Earth enhancement, taking into account key ecological pressures; their economic drivers; the distribution of drivers, impacts and constraints; and possible societal responses. For example, with climate change a key boundary metric in the planetary boundaries framework is the atmospheric concentration of carbon dioxide, and the main categories of pressures on the boundary are greenhouse gas emissions and decreasing of sinks. The main economic drivers of these pressures are combustion of fossil fuels and deforestation. These economic drivers and their impacts can be further broken down in several ways, for example geographically, into sectors and through analysis of material and energy flows. The resulting pictures of the distribution of economic drivers of climate change and their impacts then provide the basis for development societal responses—the "myriad pathways" to which Rockström et al. (2009) refer. The Driver-Pressure-State-Impact-Response (DPSIR) Framework used by the Organization for Economic Cooperation and Development (OECD) and others takes a similar approach (Häyhä et al. 2016).

Beyond creating an early-warning system based on global ecological limits, the planetary boundaries have potential to frame development of "novel and adaptive approaches to governance" (Rockström et al. 2009: 28), which could include their use in developing regulatory limits on, or incentive structures for, human activities (Rockström et al. 2009). The primary governance objective should be to establish and enforce ecological boundaries within which the human economy must function to avoid disruption and eventual collapse of interdependent ecosystems and human socio-economic systems on local, regional and global scales—that is, to orient society to living within its ecological means in a mutually enhancing human-Earth relationship.

A key challenge of governance in this context is to address global aspects of the need to reduce already critical human-induced ecological and social stresses in a socio-economic system currently oriented toward increasing them, while also confronting different cultural, ecological, economic, political and social contexts of that predicament sub-globally. Taking into account sub-global settings is also important because systems of law, policy and governance are stronger at the sub-global level that at the global level (Dearing et al. 2014; Häyhä et al. 2016). As well, the lack of an institutional architecture suitable for implementing a governance approach based on interacting planetary boundaries, as well as other significant challenges, must be overcome before planetary boundaries can have a meaningful role in governance at the global or sub-global level (Galaz et al. 2012; Häyhä et al. 2016). Making the planetary boundaries operational for policymaking and governance at different scales and across them requires explicit and distinct consideration of the biophysical (the planetary boundary processes of ecological change), socio-economic (the spatial and temporal distribution of the economic drivers of pressures on the boundaries and their impacts) and ethical (the distribution of rights and responsibilities of those who drive and those who are impacted by ecological change) dimensions (Häyhä et al. 2016). Integrating the planetary boundaries into lock-in assessments that address these and other challenges is addressed in Part IV.

The notion of bounded "safe operating space" for humanity that underlies the planetary boundaries provides part of an overall context for systems-based indicators of whether the human enterprise is supportive of a mutually enhancing human-Earth relationship within the global panarchy. Indicators derived from the planetary boundaries, including sub-global indicators (where global indicators are easily disaggregated regionally, the viability of global boundaries is doubtful or regional or local impacts may be more significant than aggregate global ones), may be helpful for determining whether the "operating space" for humanity is "safe." In addition to this human-oriented notion of safe operating space, notions such as a mutually enhancing human-Earth relationship or right relationship (Brown and Garver 2009) imply a need for additional objectives and additional indicators for Earth's social-ecological systems. In particular, previously mentioned objectives related to the flourishing of life and life systems and to interhuman, interspecies and intergenerational fairness inherent in the notion of right relationship (Brown and Garver 2009) require indicators that go beyond safe operating space for humans.

In support of applications in governance, legal systems and policy arenas, indicators based on global ecological limits could also be used to set agendas for policy-relevant law-related research, including both research aimed at improving the rigor or reducing the uncertainty of the indicators themselves and research on application of the indicators in a specific legal context. An example of the first type of research is the work analyzing the Human Development Index (HDI) against a broad range of human development criteria in order to highlight areas of development that are not well captured by the HDI (Ranis et al. 2006). Examples of the second type of research include the calculation by Quebec's Commissioner of Sustainable Development of the ecological footprint for Quebec (Lachance 2007), a study comparing the ecological footprints of conventional and organic Italian wines (Niccolucci et al. 2008) and a study of the relationship between HANPP and bird diversity in Austria (Haberl et al. 2005). Indicators can also be used as means to communicate information, for example about the indicators themselves, the objectives that underlie them or the law and policy they support (Schiller et al. 2001). A prominent example of the use of an indicator for broad communication purposes is the ubiquitous—and often misleading—use of GDP to convey information about the state of the currently dominant growth-insistent version of the economy at local, national and global levels. In applying the analytical framework described in this chapter, it is important to consider the adaptability of an indicator to uses in governance, research, communications and other arenas.

It is important as well to consider who might use this analytical framework. My intent with Part IV of this book in particular is to propose a broadly applicable, flexible method for developing strategies, priorities and timelines to transition toward a mutually enhancing human-Earth relationship. In its simplest form, the framework can be applied informally at the individual level, or in small local groups or communities. Even at that informal and small-scale level, an informed, knowledge-based, place-specific systems approach can reveal ways to confront the obstacles and opportunities that a transition to a mutually enhancing human-Earth relationship presents. In more formal, organized contexts, as with adequately funded governments, activists and private enterprises, the framework can be applied with more structure and rigor. The perspective and examples in this book are not intended to limit the potential scope of application of the proposed methodology.

Integrating planetary boundaries and the IPAT framework

To develop novel and adaptive approaches to governance, including law and legal systems that would lead to and maintain a mutually enhancing human-Earth relationship, ecological bounds on the aggregate scale of human manipulation of energy and material stocks and flows can be used to frame additional constraints and possibilities focused more directly on the socio-political dimensions of the human sphere. Aggregate global environmental impact is a function of the size of the human population worldwide, human affluence (or consumption of material and energy) and technology (the well known $I = P \times A \times T$, or IPAT, formulation, where I is impact, P is population, A is affluence or consumption and T is technology) (Ehrlich and Holdren 1972). Each of the planetary boundaries can be considered a fixed limit of the I variable, with each boundary value of I constraining the P, A and T variables; if P rises, A or T, or both, must drop. For example, suppose I represents the total carbon emissions that will safely keep the human enterprise in safe operating space, P is the human population, A is the consumption of material and energy per person and T is the amount of carbon emitted per unit of material and energy consumed. As population rises, per capita consumption (A) must decrease, the emissions per unit of consumption (T) must decrease, or both must decrease for total emissions (I) to remain unchanged. If I must decrease to return to safe conditions, as is now the case for several boundaries, the degrowth challenge becomes starker. An additional variable, ethics (E), may be introduced to this relationship to call attention to the role of ethics in making individual and collective choices regarding the other variables—making the framework $I = f(PATE)$ (Brown and Garver, 2009). As a multiplicative formula, IPAT is tautological—a way to break out impact into its drivers. As IPATE, the variables are not multiplied, but rather express variables that combine in complex ways to influence human impact (Kallis 2018: 182).

If the focus is on population, the concept of carrying capacity is a variation of the boundaries concept. The sheer size of the human population already has an enormous impact on planetary processes: humans, a species representing about 5 percent of the total animal mass of the planet (McNeill 2000), is approaching 8 billion in number in 2020 (United Nations 2019). Carrying capacity can be considered either as the maximum population that "could be sustained biophysically under given technological capabilities [or as the maximum population that could be sustained] under various social systems (and, especially, the associated patterns of resource consumption)" (Daily and Ehrlich 1992: 762). Global carrying capacity, as defined by a maximum population size, is difficult to estimate, especially as it varies geographically and by associated regional socio-economical systems. The planetary boundaries concept may help focus this effort, by allowing for an analysis of population pressures on each of the boundaries, and of possible maximum populations that will allow the boundaries to be respected given feasible options for managing the A and T variables.

Notes

1 For freshwater use, atmospheric aerosol loading, land-system change and phosphorus cycles, the 2015 update indicated regional or other sub-global limits (Steffen et al. 2015).
2 The Global Footprint Network (www.footprintnetwork.org) houses an interactive database on the correlation of ecological footprint and HDI over time for individual countries, regions and the world as a whole, https://data.footprintnetwork.org/?_ga=2.172511587.1719632431.1587341763-413128562.1515636265#/sustainableDevelopment?cn=all&yr=2016&type=BCpc,EFCpc.

References

Bosselmann, K. 2008. *The Principle of Sustainability: Transforming Law and Governance.* Burlington VT: Ashgate Publishing Co.
Boulding, K. 1975. Love and Lifeboats. Chapel Talk at Gustavs Adolphs College, St. Peter, Minnesota (November 18, 1975).
Bringezu, S., M. O'Brien and H. Schütz. 2012. Beyond biofuels: assessing global land use for domestic consumption of biomass: a conceptual and empirical contribution to sustainable management of global resources. *Land Use Policy* 29: 224–232.
Brown, P.G., and G. Garver. 2009. *Right Relationship: Building a Whole Earth Economy.* San Francisco: Berrett-Koehler.
Commoner, B., P.W. Woods, H. Eisl and K. Couchot. 2000. *Long-range Air Transport of Dioxin from North American Sources to Ecologically Vulnerable Receptors in Nunavut, Arctic Canada: Final Report to the North American Commission for Environmental Cooperation.* Montreal: Commission for Environmental Cooperation.
Daily, G.C., and P.R. Ehrlich. 1992. Population, sustainability, and Earth's carrying capacity. *BioScience* 42(10): 761–771.
Daly, H.E. 1996. *Beyond Growth.* Boston: Beacon Press.
Dearing, J.A., R. Wang, K. Zhang, J.G. Dyke, H. Haberl, M. Sarwar Hossain, P.G. Langdon, T.M. Lenton, K. Raworth, S. Brown, J. Carstensen, M.J. Cole, S.E. Cornell, T.P. Dawson, C.P. Doncaster, F. Eigenbrod, M. Flörke, E. Jeffers, A.W. Mackay, B. Mykvist and G.M. Poppy. 2014. Safe and just operating spaces for regional social-ecological systems. *Global Environmental Change* 28: 227–238.
Ehrlich, P.R., and J.P. Holdren. 1972 (May). Critique of *The Closing Circle. Bulletin of the Atomic Scientists* 16–27.

Galaz, V., F. Biermann, C. Folke, M. Nilsson and P. Olsson. 2012. Global environmental governance and planetary boundaries: an introduction. *Ecological Economics* 81: 1–3.

Garver, G. 2013. The rule of ecological law: the legal complement to degrowth economics. *Sustainability* 5(1): 316–337 [online]: www.mdpi.com/2071-1050/5/1/316/htm

Haberl, H., C. Plutzar, K-H. Erb, V. Gaube, M. Pollheimer and N.B. Schulz. 2005. Human appropriation of net primary production as determinant of avifauna diversity in Austria. *Agriculture, Ecosystems and Environment* 110: 119–131.

Halsnæs, K., P. Shukla, D. Ahuja, G. Akumu, R. Beale, J. Edmonds, C. Gollier, A. Grübler, M. Ha Duong, A. Markandya, M. McFarland, E. Nikitina, T. Sugiyama, A. Villavicencio and J. Zou. 2007. Framing issues, in B. Metz, O.R. Davidson, P.R. Bosch, R. Dave and L.A. Meyer (eds), *Climate Change 2007: Mitigation. Contribution of Working Group III to the Fourth Assessment Report of the Intergovernmental Panel on Climate Change*. New York: Cambridge University Press.

Hansen, J., M. Sato, P. Kharecha, D. Beerling, R. Berner, V. Masson-Delmotte, M. Pagani, M. Raymo, D.L. Royer and J.C. Zachos. 2008. Target atmospheric CO2: where should humanity aim? *Open Atmospheric Science Journal* 2: 217–231.

Häyhä, T., P.L. Lucas, D.P. van Vuuren, S.E. Cornell and H. Hoff. 2016. From planetary boundaries to national fair shares of the global safe operating space—how can the scales be bridged? *Global Environmental Change* 40: 60–72.

Jackson, T. 2009. *Prosperity Without Growth: Economics for a Finite Planet.* Washington, DC: Earthscan.

Kallis, G. 2018. *Degrowth.* Newcastle upon Tyne, UK: Agenda Publishing.

Lachance, R. 2007. *Rapport du vérificateur général du Québec à l'Assemblée nationale pour l'année 2007–2008*. Quebec City: Vérificateur général.

McNeill, J.R. 2000. *Something New Under the Sun: An Environmental History of the Twentieth-Century World.* New York: W.W. Norton & Co.

McNeill, J.R. 2003. Observations on the nature and culture of environmental history. *History and Theory* 42(4): 5–43.

Meadows, D., J. Randers and D. Meadows. 2004. *Limits to Growth: The 30-Year Update*. White River Junction, VT: Chelsea Green Publishing Company.

Niccolucci, V., A. Galli, J. Kitzes, R.M. Pulselli, S. Borsa and N. Marchettini. 2008. Ecological footprint analysis applied to the production of two Italian wines. *Agriculture, Ecosystems & Environment* 128(3): 162–166.

Nordhaus, T., M. Shellenberger and L. Blomqvist. 2012. *The Planetary Boundaries Hypothesis: A Review of the Evidence.* Oakland CA: The Breakthrough Institute.

Nyquist, B., A. Persson, F. Moberg, L. Persson, S. Cornell and J. Rockström. 2013. *National Environmental Performance on Planetary Boundaries: A Study for the Swedish Environmental Protection Agency.* Stockholm: Swedish Environmental Protection Agency.

Pollard, D., R. Almond, E. Duncan, M. Grooten, L. Hadeed, B. Jeffries and R. McLellan, eds. 2010. *Living Planet Report 2010*. Gland, Switzerland: World Wide Fund for Nature.

Ranis, G., F. Stewart and E. Samman. 2006. Human development: beyond the human development index. *Journal of Human Development* 7(3): 323–358.

Raworth, K. 2017. *Doughnut Economics: Seven Ways to Think Like a 21st Century Economist.* White River Junction, VT: Chelsea Green Publishing.

Rockström, J., W. Steffen, K. Noone, Å. Persson, F.S. Chapin III, E. Lambin, T.M. Lenton, M. Scheffer, C. Folke, H. Schellnhuber, B. Nykvist, C.A. De Wit, T. Hughes, S. van der Leeuw, H. Rodhe, S. Sörlin, P.K. Snyder, R. Costanza, U. Svedin, M. Falkenmark,

L. Karlberg, R.W. Corell, V.J. Fabry, J. Hansen, B. Walker, D. Liverman, K. Richardson, P. Crutzen and J. Foley. 2009. Planetary boundaries: exploring the safe operating space for humanity. *Ecology and Society* 14(2): 32 [online] URL: www.ecologyandsociety.org/vol.14/iss2/art32/

Schiller, A., C.T. Hunsaker, M.A. Kane, A.K. Wolfe, V.H. Dale, G.W. Suter, C.S. Russell, G. Pion, M.H. Jensen and V.C. Konar. 2001. Communicating ecological indicators to decision makers and the public. *Conservation Ecology* 5(1): 19.

Sen, A. 2005. Human rights and capabilities. *Journal of Human Development* 6(2): 151–166.

Steffen, W., K. Richardson, J. Rockström, S.E. Cornell, I. Fetzer, E.M. Bennett, R. Biggs, S.R. Carpenter, W. de Vries, C.A. de Wit, C. Folke, D. Gerten, J. Heinke, G.M. Mace, L.M. Persson, V. Ramanathan, B. Reyers and S. Sörlin. 2015. Planetary boundaries: guiding human development on a changing planet. *Science* 347(6223): 1259885-1–1259885-10.

United Nations. 2019. *World Population Prospects: Highlights.* New York: United Nations.

11 Eleven core features of ecological law

Ecological law is emerging because of rising concerns about the gross incapacity of contemporary law, including environmental law, to rise to the human dilemma in the Anthropocene. Four principles underlie environmental law in its contemporary form: (1) the principle that "no one has the right to cause significant harm to others," (2) the polluter pays principle, (3) the "look before you leap" principles embodied in environmental impact assessment requirements, and (4) the precautionary principle (Percival 2007: 8). Environmentalism based on these principles generally is anthropocentric and concerned, first, with protecting human health from harmful contaminants and, second, with preserving and protecting natural areas and species (Tarlock 2000: 238). Although these objectives imply some limits on human activities, the contemporary environmental law built on them has consistently defaulted to economic and political factors and has not significantly framed those limits around an integrated, systemic understanding of ecosystems and human-Earth dynamics. Nor does it effectively address the aggregate effects of pollution or biodiversity loss. Thus, the four core principles of environmental law "do not represent a system of 'transformative nature-centered rule' likely to 'tame the drive to exploit and modify all planetary life support systems'" (Percival 2007: 8). In other words, contemporary environmental law lacks principles or features that are needed to establish ecological law.

Any attempt to list the essential principles of an area of law runs the risk of missing or overstating aspects of the essence of that area of law. Rather than present a list of principles, I offer some essential features of the rule of ecological law on the understanding that there may be others. Ecological law should include at least the 11 following mutually reinforcing features (Garver 2013):

1. Humans are part of the Earth's life systems, not separate from them, and concepts that are relevant to ecological law, like ecological integrity, must include humans within flourishing life systems;
2. Ecological limits, such as planetary boundaries for safe operating space for humanity, must have primacy over economic, political and other considerations;
3. Ecological law permeates the entire legal regime in a systemic, integrated way, and is not a specialty area of the law like environmental law;

4. In light of the current ecological crisis, ecological law should focus on reducing the throughput of material and energy in the human economy;
5. Ecological law must ensure that biomass and extracted materials are obtained and used in the economy according to real needs, with minimal consideration of utilitarian desires reflected in market prices;
6. Ecological law must be global, but applied using principles of proportionality (sufficient but not excessive regulation) and subsidiarity (application of law at the lowest political tier at which the law's objective can be achieved);
7. Ecological law must ensure fair sharing of resources among present and future generations of humans and other life;
8. Ecological law must be binding and supranational, with supremacy over sub-global legal regimes as necessary;
9. A greatly expanded program of research and monitoring tied to improved understanding and continual adjustment of ecological boundaries and means for respecting them is needed to support ecological law from the global to the local level;
10. Ecological law requires precaution about crossing planetary boundaries, with both margins of safety to ensure that the boundaries are respected and complementary measures, such as ecological regeneration and restoration, to allow the Earth's life systems to flourish;
11. Ecological law must be adaptive, in recognition of the non-equilibrium nature of ecosystems and the need to get started on a comprehensive effort to constrain the economy within ecological limits despite uncertainty.

The following sections introduce these features of ecological law in more detail.

Humans are part of, not separate from, the Earth's life systems

A fundamental critique of conventional economics and law, including environmental law, is that they are grounded in the anthropocentric view that humans stand apart from, and reign over, the environment and non-human species (Berry 1999; Kysar 2002: 675–676). The most fundamental feature of ecological law is its grounding in the human-inclusive ecocentric view that humans are relational beings in a broad complex of interrelated life system. Berry (1999: 58) puts it this way:

> We might begin by recognizing that the life community, the community of all living species, including the human, is the greater reality and the greater value. The primary concern of the human community must be the preservation and enhancement of this comprehensive community, even for the sake of its own survival.

Ecological law calls for a revival of the holistic, ecocentric worldviews that were part of the conversation at the birth of contemporary environmental law but eventually ceded to a much more instrumentalist view of non-human life and life systems.

Systems-based ecological limits

As explained in Chapter 10, the goal of enclosing the social and economic spheres within the Earth's ecological limits must have primacy against other goals. The aggregate scale of, distribution within and efficiency of the economy—core themes in ecological economics—are all relevant to how well the economy and interdependent systems of law and governance maintain social and ecological well-being (Daly 1996), with scale as the constraining envelope within which distribution and efficiency must be managed (Daly and Farley 2004). Indicators of aggregate scale, for which material and energy stock and flow accounting and "planetary boundaries of safe operating space" (Rockström et al. 2009) are particularly well-suited, do not typically provide much information about distribution and efficiency. Indeed, planetary boundary researchers acknowledge that the boundaries framework "does not as yet account for regional distribution of the impact, nor of its historical patterns [and does not] take into account the deeper issues of equity and causation" (Steffen et al. 2015: 7). Nonetheless, consistent with core notions in ecological economics, staying within ecological limits takes precedence over political, economic or social considerations (Daly 1996; Rockström et al. 2009). Systems-based ecological limits stand alone as a core feature of ecological law.

Ecological law with primacy for ecological limits requires metrics of ecological limits and of social metabolism. The comprehensive framework of planetary boundaries of safe operating space for humanity (Rockström et al. 2009), supplemented with measures of social metabolism such as ecological and other footprinting (Wackernagel and Rees 1996; Häyhä et al. 2016) and human appropriation of net primary productivity (HANPP),[1] provide one promising avenue for integrating ecological limits into ecological law. Chapter 10 explained in detail how planetary boundaries are central to understanding and building on the foundation of a limits-insistent narrative. Planetary boundaries are beacons, projected from the normative rubric of "safe operating space," for guiding fundamental aspects of the human-Earth relationship—i.e. the interactions of the human enterprise, consisting of the global entirety of human endeavors, with the ecosphere, the global entirety of the biotic and abiotic elements and characteristics that make up the Earth and its atmosphere and support life on Earth (Huggett 1999). Other related boundaries are also possible, such as a limit on the global human ecological footprint that will adequately allow perpetual provisioning of the human enterprise while also maintaining acceptable levels of biodiversity (Wackernagel and Rees 1996; Ewing et al. 2010). Footprint methods also exist or are in development for all the individual planetary boundaries (Häyhä et al. 2016).

Full integration of the ecological law in legal systems

Just as ecological economics amounts to a comprehensive reorientation of conventional economics, the ecological law must become fully integrated within legal and policy infrastructures. Ecological law is not a specialty area of law as contemporary environmental law is often perceived to be, but rather a transformative alternative approach that must permeate law and policy (Kysar 2002: 682; Boyd 2004: 365). A leading expression of the need to integrate the ecological law in the entire envelope of law and policy is the Earth Charter, which states:

> To move forward we must recognize that in the midst of a magnificent diversity of cultures and life forms we are one human family and one Earth community with a common destiny. We must join together to bring forth a sustainable global society founded on respect for nature, universal human rights, economic justice, and a culture of peace. Towards this end, it is imperative that we, the peoples of Earth, declare our responsibility to one another, to the greater community of life, and to future generations.[2]

This broad and holistic vision is only possible if the principles it encapsulates are built into the entire foundation of law and governance.

Radical reduction of material and energy throughput

The global economy has grown at a supercharged rate in the industrial era of the last two centuries. The consequence is an urgent problem of the economy's scale: the economy has become too big "in its physical dimensions ... relative to the ecosystem that sustains it" (Daly and Farley 2004: 12). Rigorous analysis indicates that the human enterprise is consuming the Earth's biocapacity faster than it can regenerate and has created a rising ecological debt (Pollard et al. 2010; McLellan et al. 2014). In these circumstances, ecological law requires a radical reduction of the material and energy throughput of the economy and enforceable rules that prohibit the unnecessary use or wasting of biocapacity.

Starting in 2004, Earth system scientists have published updated graphs showing starkly the close correlation between socio-economic and Earth system trends from 1750 onward (Steffen et al. 2015). The socio-economic trend graphs show the exponential rise in global GDP, human population and various forms of socio-economic activity, such as foreign direct investment, paper consumption, large dam construction and international tourism (Steffen et al. 2015). The Earth system trend graphs show the correlative exponential rise in the ecological impacts of this socio-economic activity, such as atmospheric concentrations of greenhouse gases, ocean acidification, nutrient loading and loss of tropical forest (Steffen et al. 2015). Both sets of graphs show a sharp rise in the rate of change in trends since about 1950, a phenomenon dubbed the Great Acceleration (Steffen et al. 2015). Research showing the extent to which socio-economic activity is pushing Earth system impacts beyond planetary boundaries (Rockström et al. 2009; Steffen et al. 2015) and, since around 1970, has caused ecological overshoot in terms of ecological footprint (McLellan et al. 2014) confirms that these correlated socio-economic and Earth system trends are not merely a sign of ongoing progress, but rather indicators of an ecological crisis.

Sustainability science, which frames this crisis as a problem of the interaction between society and nature (Haberl et al. 2004: 199–200), provides several analytical tools for tracking the throughput of material and energy in the economy (Krausmann et al. 2004). Aggregate metrics of social metabolism, such as material and energy flow accounting (MEFA), metrics of the energy return on energy invested (EROEI),[3] measures of HANPP and ecological footprinting, should play an increasing role in tracking and reducing material and energy throughput at and across spatial scales.

Ecological footprint is a measure of the amount of bioproductive land (usually measured in hectares) that would be needed to process everything that humans consume and waste on a sustained basis, including the land that would be needed to take up such waste as greenhouse gases in perpetuity (Wackernagel and Rees 1996; Meadows et al. 2004). The total global human ecological footprint has exceeded the available bioproductive land on Earth more and more since about 1970 (McLellan et al. 2014). Useful footprinting methods have also been developed for all the planetary boundaries (Häyhä et al. 2016).

Material and energy flow accounting (MEFA) is used to analyze "important aspects of society–nature interaction by tracing socio-economic materials and energy flows and by assessing changes in relevant patterns and processes in ecosystems related to these flows—in other words, the 'colonization' of terrestrial ecosystems" (Haberl et al. 2004: 199–200). MEFA makes it possible to "link data and models used to analyse socio-economic systems (e.g. economic models, social statistics and indicators) to data and models used for natural systems (e.g. biosphere models, climate models, ecosystem models) in a consistent way" (Haberl et al. 2004: 203). A study of the global material flow in the twentieth century found that material extraction increased eight-fold, at an exponential rate, from 1900 to 2005, while global GDP rose by a factor of about 23 and population by a factor of four (Krausmann et al. 2009).

Clearly, an urgent challenge for ecological law is to achieve a radical reduction of material and energy throughput (Beddoe et al. 2009). "Effective governance of the global environmental commons must … proceed from a recognition of the scale of resource extraction and waste production that natural systems can accommodate in perpetuity" (von Weizsäcker et al. 2009: 223). A key additional challenge, however, is to avoid the so-called rebound effect, also known as Jevon's paradox or the Khazzoom-Brookes Postulate (von Weizsäcker 2009: 301–307). The rebound effect is the phenomenon by which increases in the efficiency of material or energy use result in increases in consumption, which often more than make up for the gains from efficiency (von Weizsäcker et al. 2009: 301). For example, the aforementioned study on global material extraction in the twentieth century makes clear that although

> an overall decline in the material intensity of the global economy, or, inversely, the increase in efficiency with which materials (and energy) are used, is a characteristic feature of the period of global industrialisation, [t]hese efficiency gains did not translate in a reduction of the materials and energy used.
>
> (Krausmann et al. 2009: 2702)

Because increases in efficiency have tended to support growth in population, affluence and consumerism, the rebound effect is a prominent feature of the current legal and socio-economic arrangements in the developed world and much of the developing world.

Many specific features of contemporary environmental law actually contribute to the situation in which overall consumption outpaces gains in the efficiency of material and energy use. For example, air pollution control in the United States often requires piecemeal implementation of the best available control technology,[4] reasonably available control technology,[5] maximum available control technology[6] and the like to individual pollution sources, which implicitly favors growth in the economic activity that is causing ecological impact over the human and ecological systems that are harmed. Similarly, Pelletier (2010: 221–222) criticizes the *LRTAP* regime governing air pollution control in Europe for implicitly protecting fossil fuel burning by imposing technological requirements that do not question the need for an ever-growing fossil fuel-based economy.

> The implicit assumptions of industrial society which permeate such regimes provide intrinsic advantage to specific actors and ideologies. Economic reliance on fossil fuels in a system premised on growth means that questions of constraint are not entertained. Instead, politically feasible reduction targets achieved through problem shifting end-of-pipe technologies are favoured over precautionary solutions, fundamental changes in production systems, and hard limits on total allowable emissions.

The principal mechanisms of contemporary environmental law carry an implicit assumption that stopping or constraining economic activity so as to meet ecological boundaries would be more devastating to the social-ecological system in which it occurs than the aggregate ecological harm from the activity would be.

The market cannot be expected to constrain material and energy throughput in the economy on its own. Rather, government standards or policies, which have been necessary historically to protect the environment from free market forces, must drive the uptake of technology that improves resource productivity, such as zero-emissions or carbon neutral technologies (Percival 2007: 30–31). To address the rebound effect, legal and policy mechanisms are also needed for directing the use of the economic gains associated with efficiency to even further reduction of the throughput of material and energy in the economy, so as to maintain sufficient resource stocks and waste processing capacity to avoid running down the Earth's ecological base (Adler 2009: 10313; von Weizsäcker et al. 2009: 314–331).

On the side of resource (or natural capital) stocks, ensuring maximum possible resource productivity and energy efficiency is achieved in social metabolism would be a good step in the direction of ecological law. For example, a wedge analysis using MEFA could be performed for different economic sectors, where the total available energy or materials would be determined based on the amount that can be extracted or made available through reuse and recycling without crossing ecological limits.[7] These capped sectoral amounts could then be distributed with market mechanisms. In essence, this would be a quota system, where resource availability is determined by ecological means, not simply price mechanisms that account for demand based on perceptions of scarcity but not on ecological limits on resource supply or waste assimilation capacity. "Trends in [resource costs, rents or prices] reflect beliefs about scarcity not necessarily the scarcity itself and it is impossible to tell from the data which it is" (Victor 2008: 54). Taxing is the other main alternative, where the objective is to adjust the price of resources so that demand stays within ecological limits on supply, as with carbon taxes (von Weizsäcker et al. 2009: 314–331).

On the waste, or pollution, side of efficiency, drastic reductions of the total output of waste from the economy are needed, either by reducing the amount of waste or pollution per unit of production or by scaling back production altogether. However, improving efficiency alone is not sufficient if overall growth in physical output is such that aggregate pollution and waste loads increase according to the rebound effect. For example, greenhouse gas "intensity," measured as the amount of greenhouse gases per unit of GDP, decreased steadily in the United States from 1990 to 2005, but aggregate greenhouse gas emissions increased (Blodgett and Parker 2010: 7). In Europe, the situation is somewhat better, in that greenhouse gas emissions decreased in the aggregate in this same period, almost certainly because of both improvements in efficiency and the displacing of production to countries outside of Europe (Blodgett and Parker 2010; Giljum and Hubacek 2001: 49).

A sector with enormous potential for improved efficiency in terms of both inputs and outputs is buildings, which at the beginning of the twenty-first century were responsible for about one-third of primary energy consumption and about 35 percent of total carbon dioxide emissions in North America (Commission for Environmental Cooperation 2008: 4–5). The green building movement focuses on using integrated design to create buildings that have low demands for materials, energy, water and waste processing and that favor reused, renewable, locally sourced and non-toxic material and energy resources to meet those demands. "A rapidly increasing market uptake of currently available and emerging advanced energy-saving technologies could result in annual reductions of 1711 megatonnes (MT) of CO2 into the atmosphere in North America by 2030, compared to a business-as-usual approach" (Commission for Environmental Cooperation 2008: 5). This reduction, which is almost as much as the carbon dioxide attributed to the entire transportation sector in the United States in 2000, would decrease total carbon dioxide emissions from the building sector in North

America in the early 2000s by about 60 percent by 2030 (Commission for Environmental Cooperation 2008: 40–47).

The example of the building sector is a reminder that gains from efficiency may be limited because of the rebound effect. From the 1950s to the early 2000s, the average single-family home in the United States grew from one hundred square meters to over two hundred square meters, even though average household size decreased from 3.67 members in 1940 to 2.62 in 2002 (Wilson and Boehland 2005). As a result of the increasing size of houses and floor area per occupant, along with the increased number of appliances, garage space and bathrooms per household, electricity consumption per household increased by 39 percent from 1970 to 2009 despite gains in efficiency (Pew Center on Global Climate Change 2009). In Canada, household energy use rose 9 percent from 1990 to 2005, during which period population increased by 17 percent and the average size of homes grew by 19 percent (Government of Canada 2007).

Where efficiency gains are insufficient to decrease aggregate use of materials and energy or production of waste, the remaining reduction in the impacts that are pushing the economy past ecological boundaries must come from constraints on production and consumption. This is where the urgent need to reduce the throughput of material and energy resources in the economy, an essential component of the rule of ecological law, confronts perhaps the most difficult dilemma of humanity's planetary crisis: the hard-wiring of growth insistence in economics and law. Through a net shift toward service sectors, some decrease in the production output of physical goods in the economy may be possible while the economy continues to grow in terms of GDP, but at some point, ecological law requires the economic growth imperative to be called into question (Daly 1996: 28; Kallis 2018). Indeed, a postulate of ecological law is that if economic growth is possible only by transgressing ecological boundaries beyond which lies catastrophic deterioration of the economy's ecological base, then an alternative to the growth paradigm is necessary (Rees 2002).

Use of material and energy only for essential needs

A close corollary in ecological law to radical reduction in social metabolism to stay within ecological limits is the requirement to limit use of material and energy to essential purposes, and not to allow their use to be determined according to utilitarian desires reflected in market prices (Martinez-Alier 2009). A broad cultural values shift, reflected in the law, is needed toward an economy that "takes no more than it needs and uses no more than it must [and] provide[s] only as much wealth as is needed for dignified, secure living" (Brown and Garver 2009: 36). Using mineral extraction as an example, Sbert (2019: 543) notes that "the current system, the reason for mining is that companies and their shareholders profit from extracting minerals for a wide array of uses, including speculation that may or may not relate to the satisfaction of basic needs." Thus, the decision of what, where, when, how and how much to mine a metal like gold is driven by and large by the price of gold in the global market—even if a great deal of gold is used for luxury items such as jewelry or held for speculative purposes (Rothe and Collins 2017). "Mining extraction cannot be based solely on economic indicators since the enormity of biophysical deterioration extends beyond conceptualizing nature purely in terms of market value or by creating compensatory measures for social and environmental degradation" (Vela-Almeida et al. 2015: 190).

Sbert (2019) incorporates this feature of law into her lens of ecological law, an analytical tool she developed for determining aspects of existing law that are consistent or inconsistent with ecological law, which she organizes around three principles: ecocentrism, ecological primacy and ecological justice. In applying the lens to an existing or proposed law, "the principle of ecological justice probes whether the law provides ethical grounding for decisions that lead to the equitable use of the planet's sustaining capacity and promotes taking only what one needs and the fair distribution of—and restraint on—wealth" (Sbert 2019: 520). Implementing this feature of ecological law will work best if communities engage in principled, participatory dialogue about what their essential needs are.

Global and sub-global rules, applied according to subsidiarity and proportionality

The foregoing sections indicate that the overall framework of ecological law must be built around ecological limits on the global aggregates of material and energy resources that the economy consumes and the wastes it produces. A central challenge is to develop legal mechanisms for distributing these global limits down to the local level in a way that respects the broad diversity of contexts, cultures, legal traditions and knowledge systems around the world. A top-down approach only will not work. If implemented on a global level, the principles of subsidiarity and proportionality that form part of the bedrock of the European Union treaties could provide strong structural support for such distribution. This architecture of distribution should also incorporate legal and policy mechanisms for enabling all humans and other living beings to flourish, built on principles of intragenerational, intergenerational and interspecies fairness. (Brown and Garver 2009: 89–91; Bosselmann 2008: 5, 85–90).

The *Treaty on European Union* (*TEU*)[8] and the *Treaty on the Functioning of the European Union (TFEU)*[9] establish subsidiarity as a core principle of governance in the European Union.[10] Article 5(3) of the *TEU* states:

> Under the principle of subsidiarity, in areas which do not fall within its exclusive competence, the Union shall act only if and in so far as the objectives of the proposed action cannot be sufficiently achieved by the Member States, either at central level or at regional and local level, but can rather, by reason of the scale or effects of the proposed action, be better achieved at Union level.

Subsidiarity has less to do with whether policymaking should be centralized or decentralized than with favoring intervention at the level at which it will be most effective for achieving policy objectives (Saunier and Meganck 2009: 259). In the United States, the federalization of environmental law that took place starting in the late 1960s looks like centralization but can be seen as an adjustment based on subsidiarity in light of evidence that state-level environmental laws were not effective enough (Engel 1997: 290). A big challenge in implementing the principle of subsidiarity is to account for a broad range of cultural, ecological and socio-political contexts, and a concomitant plurality in the way behavior is made to conform to legal limits, in fashioning a coherent global and sub-global system of ecological law.

Proportionality complements subsidiarity with the notion that governments at all levels must have sufficient capacity and authority to achieve their mandates and objectives. For example, with respect to the European Union, the *TEU* provides that "[u]nder the principle of proportionality, the content and form of Union action shall not exceed what is necessary to achieve the objectives of the Treaties." Although proportionality is sometimes viewed as a principle for limiting government, it is just as appropriately a principle for ensuring that governments are properly enabled to accomplish their work. In other words,

> [g]overnance systems and procedures at the global, national and subnational levels must have, first, the capacity to establish ecological limits on economic activity and, second, the authority to set and enforce rules that will allow human and nonhuman life and the systems they rely on to flourish for generations to come.
>
> (Brown and Garver 2009: 110)

Fairness within and among generations and species

Subsidiarity and proportionality do not necessarily ensure fairness in the allocation and distribution of the limited shares of life support capacity that human and non-human life must share. A central feature of the globally dominant economic and legal paradigm is its protection of market freedoms and property rights with a view to maximizing economic efficiency and maintaining perpetual economic growth, on the assumption that doing so will provide the greatest welfare for the greatest number of people (Daly 1996: 220; Victor 2008: 18–19, 40–41). The current economic system operationalizes utilitarian theory that is centered on the individual's pursuit of well-being, which economists assume is expressed entirely through preferences revealed in market transactions (Daly 1996: 220; Rees 2002: 251). It is reflected in the United States in constitutional notions of liberty and property rights, and in Europe in Treaty protection of the free movement of goods, of services, of capital and of people and a core commitment to economic growth. In globalized markets for goods and services, it is reflected in strong protections for individual (including corporate) market actors and preferences for privatization but relatively weak measures to protect the global commons, the poor and species other than humans from market forces (Rees 2002: 257–259). In this system, the main fairness concern is that all humans (but not other species) should have an equal possibility to consume and seek wealth, but fairness is rarely assessed in relation to initial endowments that result from inheritances, luck and actual (as opposed to theoretical) outcomes (Brown and Garver 2009: 97).

Steffen et al. (2015: 86, 91–92) show that the closely correlated socio-economic and Earth system trends during the Great Acceleration since the 1950s are not homogeneously distributed around the world. Much of the consumption and economic growth that drives global ecological change is occurring in the developed world, whereas the negative impacts are often more widespread in less developed parts of the world (Steffen et al. 2015; Häyhä et al. 2016). Researchers working on ways to make the planetary boundaries operational in policy at the national level or below, note the importance of addressing the equity dimensions of these trends (Häyhä et al. 2016). Ecological law would prioritize protecting and allowing access to the global commons—above all, the Earth's capacity to maintain life and allow flourishing—in such a way as to ensure opportunities for flourishing for a fully diverse community of living beings (Berry 1999: 28). Individual humans and artificial entities like corporations would be considered interrelational beings in a shared ecological context, and not as free agents whose quest to maximize abstract monetary wealth that can be converted into consumptive and waste-producing activities is given priority. This notion of relationship within a shared commonwealth of life provides a foundation for distribution of Earth's limited life support capacity.

Daly (1996: 220) proposes a sound principle of distribution—albeit one that should be expanded to include species other than just humans—that captures the notion of fairness that should be incorporated into ecological law: "We should strive for sufficient per capita wealth—efficiently maintained and allocated, and equitably distributed—for the maximum number of people that can be sustained over time under these conditions." This proposal rejects the utilitarian notion that humanity should "convert as much as possible of the matter/energy of the world into ourselves and our artifacts" (Daly 1996: 219), and favors striving for sufficient—not maximum—wealth for all. Equitable distribution does not require equal wealth for all, but it does imply a limited range of inequality above the level of sufficiency for all. In other words, ecological law must be open to the notion that it is possible to be too rich.

With material and energy throughput and waste production in the economy so closely correlated to GDP, a common refrain is that a moral imperative exists to allow economic growth, and hence aggregate material and energy consumption, to rise in developing countries in order to reduce poverty and increase human well-being (Gillis 2010). This is in part a question of distribution. However, the moral imperative framed this way fails to address the moral imperative to avoid a global ecological catastrophe for humanity and the commonwealth of life. The real moral imperative is to distribute access to the means for flourishing of human and other life in ways that avoid catastrophe, which must include a drastic overall reduction in material and energy throughput and waste. Economic growth may be necessary to address the situation of developing countries (Victor 2008: 18), but where growth occurs it must happen within global ecological limits. With ecological law, the notion of common but differentiated responsibilities that is weakly reflected in international law will most likely require developed countries to contract their consumption in order to ensure ecological integrity and human dignity in less developed countries (Bosselmann 2008: 59–60, 75).

Law and governance with supranational authority

Part I included an analysis of the role that strong State sovereignty has played in the lead up to the current socio-ecological dilemma. The interrelated global and sub-global dynamics underlying the pursuit of a mutually enhancing human-Earth relationship strongly imply the need for expanded supranational authority and global rules (Brown and Garver 2009). The challenge for humanity that the planetary boundaries concept frames is fundamentally global, even if its origins and manifestation are spatially and temporally diverse and complex. Thus, rules and other governance mechanisms that aim to recognize and enforce global ecological limits must be established at the global level and implemented primarily at sub-global levels so as to comport with heterogeneity in legal, cultural, political, economic and ecological contexts. The transboundary and interregional nature of the drivers and ecological manifestations of pressure on the planetary boundaries underscores the need for global rules applied locally. However, the transition to a global governance regime that promotes a mutually enhancing human-Earth relationship will almost certainly entail the emergence of more suitable power structures and political orders at all scales—such that the system for application and enforcement of rules necessary for such a regime, including the agents involved, will evolve in complex and pluralistic ways from the local to the global level.

In recognition of the challenges of addressing global challenges at the sub-global level, particularly in an international political system grounded in strong State sovereignty, Nyquist et al. (2013) conducted an analysis of "how the planetary boundaries framework offers tools and perspectives on how to work with the two-way interaction between Swedish and global environmental pressures and performance" (Nyquist et al. 2013: 9). The goal of the study was to "translate the planetary boundaries into a corresponding set of national boundaries" (Nyquist et al. 2013: 9). The study found that national performance indicators based on the planetary boundaries are best based on consumption, rather than on production within the national territory, because consumption-based indicators better account for imports that affect the planetary boundaries beyond a nation's borders. The authors also concluded that an individual nation like Sweden has limited authority to have a political impact in regulating global environmental challenges reflected in the planetary boundaries framework that affect it and to which it contributes. Yet, they also found in reviewing existing multilateral environmental agreements that "the legal infrastructure is in place to address planetary boundaries, but that the level of ambition and implementation effectiveness need to be strengthened" (Nyquist et al. 2013: 13).

The European Union provides a structural model for the kind of supranational authority that is likely needed to implement global ecological limits effectively, with three elements that stand out. First, the European treaties not only promote market freedoms, but also give European institutions the authority to adopt supranational environmental, health and safety legislation. Extending such supranational authority globally would inhibit the hiding of ecologically and socially destructive activities and interrelationships behind claims of sovereignty and limit accusations of protectionism against environmental measures that affect markets. Second, the foundational principles of subsidiarity and proportionality in the European treaties provide a strong basis for establishing global ecological limits and distributing the main responsibility for their implementation to lower jurisdictional levels. Third, a global judicial authority, which in the form of the European Court of Justice was instrumental in establishing the supranational authority of the European institutions over nation states in the areas laid out in the European treaties (Koppen 2005), is likely essential to ensuring that supranational rules are effectively implemented and enforced (Brown and Garver 2009).

The experience of Europe illustrates the emergence of supranational authority within an overall context of economic integration. Although protection of the environment was not a core objective when the Treaty of Rome established the European Community, or even mentioned in the Treaty, it emerged, with approval of the European Court of Justice, as implicit in the Community's authority to regulate the functioning of Europe's common market (Burchell and Lightfoot 2001: 34–37; Koppen 2005: 67, 72–80, 83). The federalization of environmental law in the United States under the Commerce Clause of the United States Constitution generally followed the same general logic. The chief barrier in the United States and most other countries outside of Europe to a supranational regime of ecological law is a hypersensitivity to perceived infringements of sovereignty (Johnson and Beaulieu 1996: 23–24). Importantly, however, the United States and the many countries with whom it has trade accords that include investment chapters akin to *NAFTA* Chapter 11, including Canada, Mexico, Dominican Republic, the Central American countries and Peru, have waived their sovereign immunity to allow private investors to seek binding arbitration on claims that a government has harmed its investment by violating the investment provisions. With such precedents, and with the examples of Europe and the federal system of environmental regulation in the United States, arguments against creation of a supranational regime of ecological law lose considerable force.

To oversee governance-oriented research and monitoring to address these challenges, Brown and Garver (2009) propose a Global Reserve, which would "develop and update information on the dynamics of human ecological impact, the ways in which the benefits and burdens are being allocated and distributed, and the ecological limits that must bind the economy" (Brown and Garver 2009: 112). An overarching challenge is that "[t]here are virtually no real precedents for a coherent macro-economic framework for sustainability" (Jackson 2009: 177). A Global Reserve or something like it, linked to a coherent system of enforceable supranational rules and committed to an adaptive approach that is cautious about crossing global ecological limits and incorporates lessons from experience, ongoing research and monitoring and policy innovation, could help establish such a framework.

The supranational enforcement regime in Europe and the federal system of environmental law in the United States demonstrate the value of giving non-governmental actors the ability to enforce environmental protection laws against the government or private violators. The ability of private citizens in Europe to enforce European law in the national and European courts, and of private citizens in the United States to bring citizen suits or administrative legal actions to enforce environmental laws, has undoubtedly contributed greatly to environmental progress in Europe and the United States (Boer 1995: 326; Craig and de Búrca 2008: 268–269; Tarlock 2001).

Expanded research and monitoring

Ecological law will depend on a deep scientific understanding of the global ecosystem, its subcomponents and their relationship with the human sphere. The areas in which greatly expanded research is needed fall into two categories. An adequate system of global governance will require first, more research into the Earth's life systems, their systemic behavior and thresholds and the impacts of the human enterprise on those systems; and second, ongoing research into the governance structures that are most appropriate for ecological law.

In the first category, the varying degrees of uncertainty with respect to the planetary boundaries underscores the need for greatly expanded research and monitoring in sustainability science and about Earth's life systems. For example,

> [t]here is ample evidence from local to regional scale ecosystems, such as lakes, forests, and coral reefs, that gradual changes in certain key control variables (e.g., in biodiversity, harvesting, soil quality, freshwater flows, and nutrient cycles) can trigger abrupt system state change when critical thresholds have been crossed. More research is urgently needed on the dynamics of thresholds and feedbacks that operate at continental and global scales, especially for slow changing control variables such as land use and cover, water resource use, biodiversity loss and nutrient flows.
>
> (Rockström et al. 2009: 10)

With respect to the planetary boundaries generally, "the knowledge gap is disturbing [and] there is an urgent need to identify Earth System thresholds, to analyse risks and uncertainties, and, applying a precautionary principle, to identify planetary boundaries to avoid crossing such undesired thresholds"

(Rockström et al. 2009: 28). Since 2009, planetary boundaries researchers have elaborated a broad research program to fill knowledge gaps about the planetary boundaries (Häyhä et al. 2016; van Vuuren et al. 2016; Häyhä et al. 2018).

The critical levels and loads approach to control of long-range transboundary air pollutants in Europe also illustrates the need for ongoing research and monitoring. Scientists overcame their initial concerns regarding uncertainties and simplifications of complex ecosystem dynamics used in the approach, because they agreed on the need for science-based justifications for reducing the systemic harms that air pollution causes despite uncertainties (Cresser 2000: 52; Spranger et al. 2008: 486). However, the approach spurred a great deal of research that has strengthened understanding of the relationships between pollutants and their receptor ecosystems, which in turn should lead to improvements in the approach (Cresser 2000).

More research is also needed before MEFA and related tools for analyzing socio-economic metabolism can be used in policy.

> Both MFA and EFA are, at present, restricted to accounting. Establishing MEFA compatible models that would analyse systemic interrelations between material and energy flows, land use, [human appropriation of net primary production (HANPP)], and many other important variables related to processes within social-ecological systems relevant to sustainability (e.g. net carbon flows into the atmosphere) are still in their infancy. Only such models would allow to establish scenarios and forecasts which are important for policy advice.
>
> (Haberl et al. 2004: 205)

A key problem to overcome is the current imbalance between research in support of private interests, propelled by the anticipation of financial profits, and research in support of humanity's common interest, where pooling the resources to fund research must confront competing demands for public money (Homer-Dixon 2002: 259–260). The urgency of the global ecological crisis compels a reprioritization that ensures that adequate research and monitoring is funded to enable a much better tracking of the situation of humans and our fellow Earth inhabitants with respect to the planetary boundaries.

The second category of research is oriented to the issues of public policy and governance that relate to managing the human-Earth relationship in a manner that respects global ecological boundaries and complementary principles. Some planetary boundaries researchers at the Stockholm Resilience Centre are focused on this area of research (Häyhä et al. 2016; Häyhä et al. 2018). The Earth System Governance project[11] provides another promising framework for this area of research. Earth system governance is defined in this project as

> the interrelated and increasingly integrated system of formal and informal rules, rule-making systems, and actor-networks at all levels of human

> society (from local to global) that are set up to steer societies toward preventing, mitigating, and adapting to global and local environmental change and, in particular, earth system transformation, within the normative context of sustainable development.
>
> (Biermann et al. 2009)

The research plan of the project is framed around five interlinked analytical questions, four crosscutting themes and an initial set of four "flagship activities." The analytical questions concern the architectures of governance, agency, the adaptiveness of Earth systems, accountability and legitimacy of governance systems, and allocation of and access to material and immaterial values. The crosscutting themes that relate to all the analytical questions are power, knowledge, norms and scale. The flagship activities provide broad contexts within which to examine the analytical questions and crosscutting themes, and the initial set identified for the project are the global water system, the global climate system, food systems and the global economic system. The Earth System Governance Project, in conjunction with related research efforts more clearly grounded in a critique of growth insistence, such as Research & Degrowth (R & D),[12] have considerable potential to enhance the development of ecological law.

Caution about breaching ecological limits

The need for greatly expanded research and monitoring is due to the considerable uncertainty that remains regarding the human-Earth relationship and how the human community should manage it. Ecological law requires use of the precautionary principle to guide management of the human-Earth relationship in the face of this uncertainty. In particular, caution is needed to avoid crossing planetary boundaries between safe operating space and catastrophe.

The precautionary principle was recognized in Principle 15 of the 1992 Rio Declaration,[13] which states that

> in order to protect the environment, the precautionary approach shall be widely applied by States according to their capability. Where there are threats of serious or irreversible damage, lack of full scientific certainty shall not be used as a reason for postponing cost-effective measures to prevent environmental degradation.

Other statements of the precautionary principle avoid the unfortunate reference to "cost-effective measures," such as a version adopted at the 1990 Bergen Conference that stated simply that "when confronted with serious or irreversible environmental threats, the absence of absolute scientific certainty should not serve as a pretext for delaying the adoption of measures to prevent environmental degradation" (Saunier and Meganck 2009: 229).

The precautionary principle is explicitly embedded in the bedrock of European Union environmental law. The *Treaty on the Functioning of the European*

Union provides that European Union environmental policy "shall be based on the precautionary principle."[14] Uncertainty, which triggers application of the precautionary principle, is not the same as risk. Decisions based on known risks, where probabilities of different outcomes can be compared in working toward a decision, are different from decisions in which risks are uncertain (Tallacchini 2005: S648; Sunstein 2007: 147). The precautionary principle comes into play in the increasing number of situations "in which law has to fill cognitive gaps, since scientific data prove uncertain, insufficient or susceptible to sharply diverging interpretations" (Tallacchini 2005: S648). Thus, application of the precautionary principle under the European Commission's guidance requires a prior scientific evaluation of risks and uncertainties. The Commission distinguishes prudence in regard to uncertainty in risk assessment from application of the precautionary principle, which it casts as relevant only at the risk management stage, "when scientific uncertainty precludes a full assessment of the risk and when decision-makers consider that the chosen level of environmental protection or of human, animal and plant health may be in jeopardy" (European Commission 2000: 12). The precautionary principle can be applied under this guidance to establish a presumption that a product or activity requiring government authorization is hazardous, with the burden on the proponent to prove otherwise, or to allow those concerned about the risks of a product or activity that is not a priori considered hazardous to show the dangers and risks associated with it.

The European Union applies the precautionary principle in conjunction with other core European principles, most prominently the principle of proportionality. The European Court of Justice has merged the proportionality and precautionary principles as follows:[15]

> It is true that the assessment which a Member State is required to make may reveal a high degree of scientific and practical uncertainty … . Such uncertainty, which is inseparable from the concept of precaution, influences the extent of the discretion of the Member State and thus has an impact on the means of applying the proportionality principle. In such circumstances, it must be acknowledged that a Member State may, under the precautionary principle, take protective measures without having to wait for the reality and the seriousness of those risks to be fully demonstrated … . However, the assessment of the risk cannot be based on purely hypothetical considerations … .

Questions surrounding the application of the precautionary principle in legal and policy contexts often touch on the degree of deference to be given to "science" and its findings. The degree of deference to be given to science, or whether it deserves deference at all, is strongest in a "republic of science" paradigm, in which science is seen as purely objective and "scientists form a perfect community of peers, self-governed through shared and freely discussed knowledge, without any coercive mechanisms and forms of authority other than knowledge itself" (Tallacchini 2005: S647). The law bound up with science in this idealized model gives science a strong assumption of objective truth. In reality, science is never entirely objective, and scientific uncertainty can lead to divergent views among scientists. Context and the discourse of science influence the weight of science in legal and policy settings (Tallacchini 2005: S647).

The operation of the precautionary principle gains clarity when it is considered in connection with the possibility of catastrophic harm or irretrievable loss. The concept of planetary boundaries, which are defined in relation to catastrophic systems changes, incorporates a strongly precautionary approach (Rockström et al. 2009: 7). The nature of climate change and other systemic disruptions that threaten a fundamental shift in the ecological context of the economy puts the precautionary principle in a particular context.

> [Climate change] is not an isolated externality in an otherwise perfect market system, nor a simple harm with a straightforward remedy. The traditional, often implicit assumption of a higher burden of proof for those who want change than for those who oppose it may be obsolete if the world is in fact headed rapidly for a cliff. *Something* new and different has to be done ....
>
> (Ackerman and Heinzerling 2002: 333)

The emphasis on threats of serious or irreversible environmental harm helps avoid the common criticism that the precautionary principle improperly invoked simply provides cover for a political choice to be cautious about some things but not others.

> Often people, and nations, take undue precautions against worst-case scenarios simply because they disregard the burden and risks of those precautions. But often people, and nations, neglect worst-case scenarios because they are unduly attentive to the burden imposed by precautions. It is important to look at both sides of the Ledger.
>
> (Sunstein 2007: 7–8)

To avoid a state of paralysis in which the precautionary principle prevents any decisions because of risks on both sides of the ledger, a system is needed for ordering risks and accounting for uncertainty.

In the case of the risk of catastrophic harm, the precautionary principle can properly order public choice among opposing risks in the face of uncertainty. Sunstein (2007: 167–168) proposes the Catastrophic Harm Precautionary Principle:

> In deciding whether to eliminate the worst-case scenario under circumstances of uncertainty, regulators should consider the losses imposed by eliminating that scenario, and the size of the difference between the worst-case scenario under one course of action and the worst-case scenario under alternative courses of action. If the worst-case scenario under one course of action is much worse than the worst-case scenario under another course

> of action, and if it is not extraordinarily burdensome to take the course of action that eliminates the worst-case scenario, regulators should take that course of action. But if the worst-case scenario under one course of action is not much worse than the worst-case scenario under another course of action, and if it is extraordinarily burdensome to take the course of action that eliminates the worst-case scenario, regulators should not take that course of action.

This formulation may be applied in ways that avoid undervaluation of future benefits of imposing immediate measures to avoid catastrophe (the discounting problem), that account for trade-offs between different (and sometimes opposing) risks of catastrophe, and that avoid unreasonable risk aversion in the face of uncertainty (Sunstein 2007: 118–175). Thus, a precautionary approach might call for elimination of the catastrophic consequences of climate change even at the cost of a 30 percent reduction in living standards, but only if the probability of those consequences rises above some threshold "for extremely significant expenditures in response to a catastrophic risk under conditions of uncertainty" (Sunstein 2007: 171).

Sunstein (2007: 177) also introduces the Irreversible Harm Precautionary Principle, which provides that "[s]pecial steps should be taken to avoid irreversible harms, through precautions that go well beyond those that would be taken if irreversibility were not a problem." This version of the precautionary principle is grounded in valuing and preserving options and flexibility that may be irreversibly lost if certain decisions are taken, and in valuing deferral of decisions causing irreversible harm if it appears possible to reduce uncertainty about the nature and degree of harm (Sunstein 2007: 180–182). "The key point is that uncertainty and irreversibility should lead to a sequential decision-making process. If better information will emerge, regulators might seek an approach that preserves greater flexibility" (Sunstein 2007: 182).

Sunstein (2007: 184–186) clarifies that irreversibility should involve harms that "rise to a certain level of magnitude" and—to account for the possibility that avoiding one irreversible loss may cause other irreversible losses—care should be taken to assess the extent of irreversibility related to alternative choices. For example, some argue that it is better to wait for better information before deciding whether and how to address climate change, given the possibility of irreversible decisions involved in mitigating it—although "we have good reason to believe that the irreversible losses associated with climate change do indeed justify the irreversible losses associated with greater investment in emissions reductions, worldwide" (Sunstein 2007: 187).

Different things that might be lost or gained in an irreversible decision might be incommensurable, such that, for example, monetary compensation will never be sufficient or correctly calculated for the loss of a unique, pristine area or a rare species (Sunstein 2007: 188). The notion of incommensurability is important for understanding the fundamental problem with the view that the future losses from transgressing planetary boundaries such as climate change or biodiversity loss can be reduced to a monetary value, discounted to present value and compared against the anticipated growth in alternative investments of the money that would have to be spent now to avoid those future losses (Martinez-Alier et al. 1998).

Even limited to catastrophic or irreversible environmental harms, one might try to avoid application of the precautionary principle to avoid ecological catastrophe by characterizing the economic or social implications of being ecologically cautious as equally or more catastrophic. For example, catastrophic loss of employment or negative effects on economic growth have been invoked to resist action to regulate greenhouse gases in the United States—often by climate skeptics or people who rely on high discounting of future consequences of climate change (Inhofe 2007; Oreskes and Conway 2010). However, these are typically false comparisons, in that they rely on incommensurable forms of catastrophe. The economic catastrophes that are typically contrasted with catastrophes associated with the planetary boundaries, in particular climate change catastrophes, ignore the possibility of alternative economic scenarios that might avoid economic catastrophe, such as innovations in access to and distribution of the means of well-being (Victor 2008: 191–224). Further, the supposed economic catastrophes resulting from mitigation of climate change must be compared to the projected economic catastrophes of not mitigating climate change.

With ecological law, Sunstein's guidelines for applying the precautionary principle when faced with catastrophic harm or irreversible loss lay a possible foundation for managing collective human behavior and activities so as to steer clear of planetary boundaries. However, in applying his rules, it is important to respect the incommensurability of many ecological values with monetary indicators of value, and to apply a strong assumption that many so-called ecosystem services have infinite monetary value and must be protected, preserved and enhanced using non-market mechanisms.

Adaptive approaches

Building off the need for expanded research and monitoring and a precautionary approach giving primal importance to global ecological boundaries, ecological law should take an adaptive approach. An adaptive approach is warranted for two main reasons.

First, in order to exercise caution about crossing planetary and sub-global ecological boundaries, ecological constraints on the human enterprise must be integrated into the global legal and policy structure despite inevitable uncertainties, which will persist in some form or another. An adaptive approach allows mechanisms to be put in place to fend off catastrophe and adjusted as research and experience fill gaps in knowledge about Earth systems and about governance of the human-Earth relationship. The adaptiveness called for applies both to the response to evolving scientific understandings and to the mechanisms and institutional arrangements in which to apply them. Among other things, an adaptive approach must consistently update and refine the establishment and fulfillment of common but differentiated responsibilities of developed and developing countries.

Second, adaptation is needed in recognition of the non-equilibrium nature and complexity of ecosystems. A key development in the science of ecology and its relationship to law in the last few decades has been the switch from an equilibrium view of nature, in which ecosystems were assumed to have an ideal natural state, to a non-equilibrium view, in which ecosystems are now seen as constantly evolving, often in stochastic and non-linear ways (Tarlock 1996).

The need for adaptiveness extends to the institutions needed to support ecological law. Law and policy tend to favor finality and certainty and resist change (Tarlock 1996: 202; Daniels 2007), and thus transition to adaptive approaches for law and legal institutions will be difficult. Institutions designed to solve and avoid environmental problems are essential, but they can stagnate, fail to meet the objectives for which they were established and yet stay in place because of bureaucratic inertia (Daniels 2007: 565). The institutions established around environmental provisions in trade and investment agreements meet this description, as explained in detail in Chapter 14.

Notes

1 HANPP is a measure of the amount of natural biomass production that humans either harvest or appropriate by other means (Haberl et al. 2007).
2 The Earth Charter Initiative, *The Earth Charter* (2000) [online]: www.earthcharterinaction.org/content/pages/Read-the-Charter.html Preamble.
3 EROEI is the measure of the amount of energy obtained per unit of energy invested to obtain it.
4 *Clean Air Act* § 169, 42 U.S.C. § 7479 (2006).
5 *Clean Air Act* § 172(c)(1), 42 U.S.C. § 7502(c)(1) (2006).
6 *Clean Air Act* § 112(g)(2), 42 U.S.C. § 7412(g)(2) (2006).
7 By analogy, Pacala and Socolow (2004) propose "stabilization wedges" in which the reductions in greenhouse gases needed to stabilize the level of greenhouse gases in the atmosphere are distributed among wedges representing different economic sectors, such as vehicles, buildings and energy shifts.
8 *Consolidated version of the Treaty on European Union*, [2008] O.J. C 115/13, *incorporating changes made by Treaty of Lisbon Amending the Treaty on European Union and the Treaty Establishing the European Community*, 13 December 2007, O.J. [2007] C 306/01 (entered into force 1 December 2009).
9 *Consolidated version of the Treaty on the Functioning of the European Union,* [2008] O.J. C 115/47, *incorporating changes made by Treaty of Lisbon Amending the Treaty on European Union and the Treaty Establishing the European Community*, 13 December 2007, O.J. [2007] C 306/01 (entered into force 1 December 2009).
10 See also *Protocol 2 On the Application of the Principles of Subsidiarity and Proportionality*. The Protocol establishes procedures for consultations on draft legislative proposals with regard to subsidiarity and proportionality and requires European Union institutions to justify draft legislative proposals with regard to the principles of subsidiarity and proportionality.
11 Earth System Governance Project [online]: www.earthsystemgovernance.org/.
12 Research & Degrowth (R & D) [online]: https://degrowth.org.
13 UNEP, *Rio Declaration on Environment and Development* (1992), Principle 15.
14 *TFEU* art. 191.
15 *Commission v. France*, C-333/08, [2010] E.C.R.—at para. 91 (citations omitted).

References

Ackerman, F., and L. Heinzerling. 2002. Pricing the priceless: cost benefit analysis of environmental protection. *University of Pennsylvania Law Review* 150(5): 1553–1584.

Adler, R.W. 2009. Freshwater: sustaining use by protecting ecosystems. *Environmental Law Reporter News & Analysis* 39: 10309–10315.

Beddoe, R., R. Costanza, J. Farley, E. Garza, J. Kent, I. Kubiszewski, L. Martinez, T. McCowen, K. Murphy, N. Myers, Z. Ogden, K. Stapleton and J. Woodward. 2009. Overcoming systemic roadblocks to sustainability: the evolutionary redesign of worldviews, institutions, and technologies. *Proceedings of the National Academy of Science* 106(8): 2483–2489.

Berry, T. 1999. *The Great Work: Our Way Into the Future*. New York: Three Rivers Press.

Biermann, F.M., M. Betsill, J. Gupta, N. Kanie, L. Lebel, D. Liverman, H. Schroeder and B. Siebenhüner. 2009. *Earth System Governance: People, Places and the Planet.* Bonn: The Earth System Governance Project.

Blodgett, J., and L. Parker. 2010. *Greenhouse Gas Emission Drivers: Population, Economic Development and Growth, and Energy Use.* Washington, DC: Congressional Research Service.

Boer, B. 1995. Institutionalising ecologically sustainable development: the rules of national, state, and local governments in translating grand strategy into action. *Willamette Law Review* 31: 307–358.

Bosselmann, K. 2008. *The Principle of Sustainability: Transforming Law and Governance.* Burlington VT: Ashgate Publishing Co.

Boyd, D.R. 2004. Sustainability law: (R)Evolutionary directions for the future of environmental law. *Journal of Environmental Law and Practice* 14: 357–384.

Brown, P.G., and G. Garver. 2009. *Right Relationship: Building a Whole Earth Economy.* San Francisco: Berrett-Koehler.

Burchell, J., and S. Lightfoot. 2001. *The Greening of the European Union?* London: Sheffield Academic Press.

Commission for Environmental Cooperation (CEC). 2008. *Green Building in North America: Opportunities and Challenges.* Montreal: Commission for Environmental Cooperation.

Craig, P., and G. de Búrca. 2008. *EU Law: Text, Cases, and Materials.* Oxford: Oxford University Press.

Cresser, M.S. 2000. The critical loads concept: milestone or millstone for the new millennium? *The Science of the Total Environment* 249: 51–62.

Daly, H.E. 1996. *Beyond Growth.* Boston: Beacon Press.

Daly, H.E., and J. Farley. 2004. *Ecological Economics: Principles and Applications.* Washington, DC: Island Press.

Daniels, B. 2007. Emerging commons and tragic institutions. *Environmental Law* 37(3): 515–571.

Engel, K.H. 1997. State environmental standard-setting: is there a "race" and is it "to the bottom"? *Hastings Law Journal* 48: 271–398.

European Commission. 2000. Communication from the Commission on the precautionary principle. Brussels.

Ewing, B., D. Moore, S. Goldfinger, A. Oursler, A. Reed and M. Wackernagel. 2010. *The Ecological Footprint Atlas 2010.* Oakland CA: Global Footprint Network.

Garver, G. 2013. The rule of ecological law: the legal complement to degrowth economics. *Sustainability* 5(1): 316–337 [online]: www.mdpi.com/2071-1050/5/1/316/htm

Giljum, S., and K. Hubacek. 2001. *International Trade, Material Flows and Land Use: Developing a Physical Trade Balance for the European Union.* Interim Report IR-01–159. Laxenburg, Austria: International Institute for Applied Systems Analysis.

Gillis, J. 2010. A scientist, his work and a climate reckoning. *The New York Times* (21 December 2010) [online]: www.nytimes.com/2010/12/22/science/earth/22carbon.html?pagewanted=5&hp.

Government of Canada. 2007. Sector Sustainability Tables, Economic Scan of Canada's Energy Sector [online]: www.tdds-sst.gc.ca/default.asp?lang=En&n=AADC6287-1&offset=3&toc=show.

Haberl, H., K-H. Erb, F. Krausmann, V. Gaube, A. Bondeau, C. Plutzar, S. Gingrich, W. Lucht and M. Fischer-Kowalski. 2007. Quantifying and mapping the human appropriation of net primary production in earth's terrestrial ecosystems. *Proceedings of the National Academy of Science* 104(31): 12942–12947.

Haberl, H., N.B. Schulz, C. Plutzar, K.-H. Erb, F. Krausmann, W. Loibl, D. Moser, N. Sauberer, H. Weisz, H.G. Zechmeister and P. Zulka. 2004. Human appropriation of net primary production and species diversity in agricultural landscapes. *Agriculture, Ecosystems and Environment* 102: 213–218.

Häyhä, T., Cornell, S.E., Hoff, H., Lucas, P., van Vuuren, D., 2018. *Operationalizing the Concept of a Safe Operating Space at the EU Level—First Steps and Explorations.* Stockholm Resilience Centre Technical Report, prepared in collaboration with Stockholm Environment Institute (SEI) and PBL Netherlands Environmental Assessment Agency. Stockholm: Stockholm Resilience Centre.

Häyhä, T., P.L. Lucas, D.P. van Vuuren, S.E. Cornell and H. Hoff. 2016. From planetary boundaries to national fair shares of the global safe operating space—how can the scales be bridged? *Global Environmental Change* 40: 60–72.

Homer-Dixon, T. 2002. *The Ingenuity Gap: Facing the Economic, Environmental, and Other Challenges of an Increasingly Complex and Unpredictable World.* New York: Vintage Books.

Huggett, R.J. 1999. Ecosphere, biosphere, or Gaia? What to call the global ecosystem. *Global Ecology and Biogeography* 8(6): 425–431.

Inhofe, S.J. 2007. Floor Speech: Inhofe, Boxer Debate Global Warming on Senate Floor, *Press Release* (29 October 2007) [online]: http://epw.senate.gov/public/index.cfm?FuseAction=Minority.PressReleases&ContentRecord_id=eddb00e1-802a-23ad-4b39-b70900f52b98&Region_id=&Issue_id=.

Jackson, T. 2009. *Prosperity Without Growth: Economics for a Finite Planet.* Washington, DC: Earthscan.

Johnson, P-M., and A. Beaulieu. 1996. *The Environment and NAFTA: Understanding and Implementing the New Continental Law.* Washington, DC: Island Press.

Kallis, G. 2018. *Degrowth.* Newcastle upon Tyne, UK: Agenda Publishing.

Koppen, I.J. 2005. The Role of the European Court of Justice, in A. Jordan, ed., *Environmental Policy in the European Union, Second Edition.* Sterling, VA: Earthscan.

Krausmann, F., H. Haberl, K.-H. Erb and M. Wackernagel. 2004. Resource flows and land use in Austria 1950–2000: using the MEFA framework to monitor society–nature interaction for sustainability. *Land Use Policy* 21: 215–230.

Krausmann, F., S. Gingrich, N. Eisenmenger, K.H. Erb, H. Haberl and M. Fischer-Kowalski. 2009. Growth in global materials use, GDP and population during the twentieth century. *Ecological Economics* 68: 2696–2705.

Kysar, D.A. 2002. Law, environment, and vision. *Northwestern University Law Review* 97: 675–729.

Martinez-Alier, J. 2009. Socially sustainable economic de-growth. *Development and Change* 40(6): 1099–1119.

Martinez-Alier, J., G. Munda and J. O'Neill. 1998. Weak comparability of values as a foundation for ecological economics. *Ecological Economics* 26: 277–286.

McLellan, R., L. Iyengar, B. Jeffries and N. Oerlemans, eds. 2014. *Living Planet Report 2014: Species and Spaces, People and Places.* Gland, Switzerland: WWF.

Meadows, D., J. Randers and D. Meadows. 2004. *Limits to Growth: The 30-Year Update.* White River Junction, VT: Chelsea Green Publishing Company.

Nyquist, B., A. Persson, F. Moberg, L. Persson, S. Cornell and J. Rockström. 2013. *National Environmental Performance on Planetary Boundaries: A Study for the Swedish Environmental Protection Agency.* Stockholm: Swedish Environmental Protection Agency.

Oreskes, N., and E.M. Conway. 2010. *Merchants of Doubt.* New York: Bloomsbury Press.

Pacala, S., and R. Socolow. 2004. Stabilization wedges: solving the climate problem for the next 50 years with current technologies. *Science* 305: 968–972.

Pelletier, N. 2010. Of laws and limits: an ecological economic perspective on redressing the failure of contemporary global environmental governance. *Global Environmental Change* 20: 220–228.

Percival, R.V. 2007. Environmental law in the twenty-first century. *Virginia Environmental Law Journal* 25: 1–35.

Pew Center on Global Climate Change. 2009. *Residential Building End-Use Efficiency* [online]: www.pewclimate.org/technology/factsheet/ResidentialBuildingEnd-Use.

Pollard, D., R. Almond, E. Duncan, M. Grooten, L. Hadeed, B. Jeffries and R. McLellan, eds. 2010. *Living Planet Report 2010.* Gland, Switzerland: World Wide Fund for Nature.

Rees, W.E. 2002. Globalization and sustainability: conflict or convergence? *Bulletin of Science, Technology & Society* 22.

Rockström, J., W. Steffen, K. Noone, Å. Persson, F.S. Chapin III, E. Lambin, T.M. Lenton, M. Scheffer, C. Folke, H. Schellnhuber, B. Nykvist, C.A. De Wit, T. Hughes, S. van der Leeuw, H. Rodhe, S. Sörlin, P.K. Snyder, R. Costanza, U. Svedin, M. Falkenmark, L. Karlberg, R.W. Corell, V.J. Fabry, J. Hansen, B. Walker, D. Liverman, K. Richardson, P. Crutzen and J. Foley. 2009. Planetary boundaries: exploring the safe operating space for humanity. *Ecology and Society* 14(2): 32 [online] URL: www.ecologyandsociety.org/vol.14/iss2/art32/

Rothe, D.L., and V.E. Collins. 2017. International Financial Institutions as Facilitators of Environmental Crimes, Chapter 6 in T. Bergin and E. Orlando, eds., *Forging a Socio-Legal Approach to Environmental Harms.* New York: Routledge.

Saunier, R.E., and R.A. Meganck. 2009. *Dictionary and Introduction to Global Environmental Governance, Second Edition.* London: Earthscan.

Sbert, C. 2019. El Salvador's mining ban and mining in Ontario's ring of fire from the lens of ecological law. *Vermont Law Review* 43(3): 517–547.

Spranger, T., J-P. Hettelingh, J. Slootweg and M. Posch. 2008. Modelling and mapping long-term risks due to reactive nitrogen effects: an overview of LRTAP convention activities. *Environmental Pollution* 154: 482–487.

Steffen, W., K. Richardson, J. Rockström, S.E. Cornell, I. Fetzer, E.M. Bennett, R. Biggs, S.R. Carpenter, W. de Vries, C.A. de Wit, C. Folke, D. Gerten, J. Heinke, G.M. Mace, L.M. Persson, V. Ramanathan, B. Reyers and S. Sörlin. 2015. Planetary boundaries: guiding human development on a changing planet. *Science* 347(6223): 1259885-1–1259885-10.

Sunstein, C.R. 2007. *Worst-Case Scenarios.* Cambridge MA: Harvard University Press.

Tallacchini, M. 2005. Before and beyond the precautionary principle: epistemology of uncertainty in science and law. *Toxicology and Applied Pharmacology* 207: S645–S651.

Tarlock, A.D. 1996. Environmental law: ethics or science. *Duke Environmental Law & Policy Forum* 7: 193–223.

Tarlock, A.D. 2000. The future of environmental "rule of law" litigation. *Pace Environmental Law Review* 17: 237–272.

Tarlock, A.D. 2001. The future of environmental "rule of law" litigation and there is one. *Pace Environmental Law Review* 19: 611–617.

van Vuuren, D.P., P.L. Lucas, T. Häyhä, S.E. Cornell and M. Stafford-Smith. 2016. Horses for courses: analytical tools to explore planetary boundaries. *Earth System Dynamics* 7: 267–279.

Vela-Almeida, D., G. Brooks and N. Kosoy. 2015. Setting the limits to extraction: a biophysical approach to mining activities. *Ecological Economics* 119: 189–196.

Victor, P. 2008. *Managing Without Growth.* Cheltenham, UK: Edward Elgar.

von Weizsäcker, E., K. Hargroves, M.H. Smith, C. Desha and P. Stasinopoulos. 2009. *Factor Five: Transforming the Global Economy through 80% Improvements in Resource Productivity.* London: Earthscan.

Wackernagel, M., and W.E. Rees. 1996. *Our Ecological Footprint: Reducing Human Impact on the Earth.* Gabriola Island, BC: New Society Publishers.

Wilson, A., and J. Boehland. 2005. *Small is Beautiful: U.S. House Size, Resource Use, and the Environment* [online]: www.greenbiz.com/news/2005/07/12/small-beautiful-us-house-size-resource-use-and-environment>.

12 Ecological integrity and attachment to place in ecological law

Two important concepts are embedded throughout the 11 features of ecological law described in the previous chapter and warrant special attention. The first is ecological integrity, a term used frequently but not always consistently in the environmental arena. Ecological integrity resonates with the idea of a mutually enhancing human-Earth relationship, so its meaning deserves clarification. The second is attachment to place, a concept also central to the idea of a mutually enhancing human-Earth relationship. Attachment to place is at the heart of many Indigenous legal traditions, place-based governance systems that have emerged in ways that sustain human reliance on ecosystems over long time periods and successful examples of ecological and eco-cultural restoration.

Overview of challenges in defining ecological integrity

Many international agreements or soft law instruments refer to ecological integrity as an overarching, or at least significant, objective (Kim and Bosselmann 2013). One example is the Earth Charter, which was adopted in 2000 as an alternative to the Rio Declaration on Environment and Development (including the Rio Principles) adopted at the 1992 United Nations Conference on Environment and Development in Rio (also known as the Earth Summit). The Earth Charter sets out one of its four core sets of principles as follows: "Protect and restore the integrity of Earth's ecological systems, with special concern for biological diversity and the natural processes that sustain life" (Earth Charter 2020). The Earth Charter broadly aligns ecological integrity with prevention of environmental harm using a precautionary approach; adoption of "patterns of production, consumption, and reproduction that safeguard Earth's regenerative capacities, human rights, and community well-being[; and] [a]dvance[ment of] the study of ecological sustainability and promot[ing] the open exchange and wide application of the knowledge acquired" (Earth Charter 2020). The outcome document at the Rio+20 conference on sustainable development, *The Future We Want*, called for "holistic and integrated approaches to sustainable development which will guide humanity to live in harmony with nature and lead to efforts to restore the health and integrity of the Earth's ecosystem" (United Nations 2012).

Ecological integrity and related notions of integrity are also integral to other broad proposals for an overarching ethic for law in the Anthropocene. Leopold's enduring environmental ethic is grounded in enhancing the integrity of the biotic community (Leopold 1949). The restatement of this ethic by Brown and Garver (2009) reiterates the moral charge to preserve and enhance the integrity of the commonwealth of life in order to ensure "right relationship" within the Earth system. Right relationship refers to a "guidance system for functioning in harmony with scientific reality and enduring ethical traditions" (Brown and Garver 2009: 4). The normative frame of safe operating space for humanity that underlies planetary boundaries proposals includes consideration of biosphere integrity (Steffen et al. 2015). Because of its prominence in existing instruments and other normative proposals, Kim and Bosselmann (2013) contend that, with appropriate refinement and clarity, ecological integrity could emerge as the supreme objective of law in the Anthropocene.

Ecological integrity—and related concepts such as biological integrity (Angermeier and Karr 1994), biosphere integrity (Steffen et al. 2015), integrity of the biotic community (Leopold 1949), biome integrity (Mace et al. 2014) and ecosystem integrity (De Leo and Levin 1997)—tends to connote notions of health, wholeness and naturalness (Angermeier and Karr 1994; Westra et al. 2000; Ordonez and Duinker 2012; Bridgewater et al. 2014). Ecological or biological integrity is usually defined with reference to pristine (or relatively pristine) ecosystems, such as wilderness areas or other protected areas that are far from human settlements and relatively free of anthropogenic impacts (Ordonez and Duinker 2012)—for example, "wild nature ... that is virtually unchanged by human presence or activities" (Westra et al. 2000: 20). So defined, ecological integrity potentially can be assessed by comparing an ecosystem's structure and function with benchmark conditions in a comparable ecosystem that is considered relatively free of human impacts, taking into account natural variability (Angermeier and Karr 1994; Westra et al. 2000; Ordonez and Duinker 2012).

The Anthropocene concept has significant implications for the meaning of ecological integrity. The Anthropocene underscores the near certainty that the global reach of human impacts on Earth's life systems is such that ecosystems completely free of anthropogenic impacts no longer exist—and probably that widespread human impacts on "wild nature" have occurred since Paleolithic times (Ellis and Ramankutty 2008; Minteer 2012; Ellis 2013; Foley et al. 2013; Ruddiman et al. 2014; Lewis and Maslin 2015). Taking into account the full history of human impacts on ecosystems from the local to global scale, ecological integrity defined strictly in reference to pristine nature free of humans is no longer possible—and likely never was. Ultimately, at the global scale, a purist definition of ecological integrity implies the elimination of the human species, in line with the "cult of wilderness" paradigm (Merchant 2003). Moreover, so defined, it runs the risk of perpetuating the human-nature divide that it aims to overcome (M'Gonigle 2008).

Accounting for historical human impacts in defining ecological integrity

To account for the history of human-Earth relations, a more coherent understanding of ecological integrity is needed. To serve as the foundation for an overarching, global objective that can be applied at the scale of landscapes or the entire Earth, ecological integrity must encompass not only the most pristine wild ecosystems, but also dense human settlements and other areas in which humans or their impacts have significantly transformed the evolutionary trajectory of the pre-human or an imagined human-free ecosystem (Ellis and Ramankutty 2008). Perhaps in recognition of the lack of stark delineations, benchmarks of ecological integrity have sometimes been tied to ecosystems that are "virtually" pristine or "relatively" free of anthropogenic impacts (Westra et al. 2000). Yet, by departing from a pure notion of integrity, these qualifiers highlight the need for additional criteria. Establishing a coherent and universal meaning of ecological integrity that renders it useful for addressing the multifold challenges of the Anthropocene within a comprehensive global legal context requires sorting out several complexities.

To begin, although concepts like health and integrity have a general resonance and appeal, defining them precisely and objectively as opposed to qualitatively and subjectively is elusive, if not impossible (Wicklum and Davies 1995; Goldberg et al. 2015). Indeed, "[h]ealth and integrity are not inherent properties of ecosystems" (Wicklum and Davies 1995: 999). Attempts to define integrity for management purposes have tended to (1) tie it to static conditions that are at odds with change processes in ecosystems, (2) rely too heavily on an indefensible presumption of absence from human impact, (3) wrongly assume that ecosystems have homeostatic processes that maintain self-organization and serve to drive them toward unchanging end states and (4) imbue it with value-laden judgments regarding the desired state or trajectory of the system (Wicklum and Davies 1995).

The notion of ecosystem health has been distinguished from ecological integrity. Integrity is usually associated with wild systems that are relatively pristine and free of human interference in perpetuity. By contrast, health is associated with the near-term ability of a system—even in an altered or managed state, as with an agricultural monoculture—to fulfill a desired function reliably (Westra 2011). Further, integrity is associated with maintaining a system's maximum range of evolutionary pathways, for example by maintaining a certain level of species diversity and abundance, whereas health may exist with respect to a smaller range of desired options (Westra 2011). Yet, even if ecological integrity is distinguished from the more elusive concept of ecological health, identifying objective criteria for integrity consistent with a mutually enhancing human-Earth relationship remains a challenge.

As already alluded to, any definition of ecological integrity must account for the continual evolution that ecosystems undergo as complex adaptive systems. This can render reference to any static set of conditions or benchmarks inappropriate (Wiklum and Davies 1995; Bridgewater et al. 2014). Concerns in this regard (Wiklum and Davies 1995) may be overstated, in that ecological integrity has more recently been aligned with evolutionary trajectories, rather than a static set of optimum conditions, consistent with contemporary developments in ecology (Higgs 2003). These developments mean that integrity must be assessed according to systems' long-term cycles and functions (Gunderson and Holling 2002; Brown and Garver 2009). Yet, aligning ecological integrity with the moving targets of evolutionary trajectories introduces additional complexities and challenges.

One such challenge is the question of both spatial and temporal scale, which must be part of any coherent understanding of ecological integrity. The spatial scale at which ecological integrity can be considered ranges from the global to the very local. The ecosystem structure and functioning that must be accounted for, including all relevant processes and interacting elements, varies significantly across scales horizontally and from one scale to another vertically (Bridgewater et al. 2014). Researchers seeking to define ecological integrity implicitly acknowledge that, scaled to the global level, full ecological integrity for the global ecosystem would only be possible without humans—or at least modern humans (Westra et al. 2000; Westra 2011). The most pristine ecosystems are part of higher-scale landscapes and ecosystems, and ultimately the entire ecosphere (Westra et al. 2000). The anthropogenic impacts or alterations in those larger landscapes and ecosystems at some point become significant, compromising ecological integrity (if defined with reference to pristine nature) at least to some degree. For example, a prominent advocate of giving primacy to ecological integrity has noted that "[n]o completely evolutionary processes may persist in an urban area (although reproduction of some species continues), but successful work has been done to show that such areas *can* be made compatible with integrity" (Westra 2011: 332). Yet, the criteria for compatibility in such a case, which must account for all the underlying systems complexities, remain elusive. References to biophysical conditions of the Holocene reflecting global ecological integrity, because they "preceded the anthropogenic global environmental change that began with the Industrial Revolution" (Kim and Bosselmann 2013: 306), point in the right direction but may still be insufficient in light of global anthropogenic impacts that occurred or began prior to the Industrial Revolution.

Temporal scale presents other complexities. Any practicable definition of a mutually enhancing human-Earth relationship must account for temporal trends within and across spatial scales in a vast, integrated complex of social and ecological systems—that is, the global panarchy (Gunderson and Holling 2002). The implications are expansive. Complex adaptive systems inevitably evolve, typically with cyclical changes in the arrangements of structure and function (Gunderson and Holling 2002). As well, the path dependence of systems limits the scope of possible futures given past systems conditions, and the past-future divide that defines path dependence is constantly advancing. The resulting irreducible uncertainty in the evolution of feedbacks and other systems dynamics across temporal scales, from the local to the global, makes it impossible to predict exactly the performance of the human-Earth system over time. Thus, giving priority to an enduring human-Earth partnership requires a broad, coordinated and adaptive system of law and governance from the local to the global level, taking into account temporal factors.

Although humanity might endeavor to maintain or restore the highest levels of ecological integrity possible, the significant existing human impacts associated with the Anthropocene make full ecological integrity (defined with reference to pristine nature) everywhere on Earth (the global scale)—indeed, anywhere on Earth (the most local scale)—impossible, even in the best of circumstances. In the Anthropocene, a mutually enhancing human-Earth relationship, and any practicable concept of ecological integrity, must allow for some degree of human impacts on the ecosystems in which people are embedded. Studies confirming earlier work showing human impacts on Earth systems dating back to Neolithic or Paleolithic times (Ellis and Ramankutty 2008; Ellis 2013; Foley et al. 2013; Ruddiman 2013) suggest that for much of human history, at different times and in different places, such a relationship may well have existed—and therefore, that environmental history can help reveal factors that either help or impede it (McNeill 2003; Diamond 2005).

Planetary boundaries research may also lead to an understanding of ecological or biosphere integrity that accounts for the complexities noted above and that is broadly oriented toward a mutually enhancing human-Earth relationship. The planetary boundaries discussed at length in Chapter 10, are conceived as limits beyond which humans face "the risk of deleterious or even catastrophic environmental change at continental to global scales" (Rockström et al. 2009: 1). A key planetary boundary—one of two, along with climate change, that researchers refer to as core boundaries because of their importance for the functioning of the Earth system—is biosphere integrity (Steffen et al. 2015). Biosphere integrity has two components. The first, currently assessed according to the rate of species extinction, has to do with maintaining genetic diversity so as to allow living and non-living components of the Earth system to evolve "in the most resilient way possible" (Steffen et al. 2015: 1259855-5). The second component "aims to capture the role of the biosphere in Earth-system functioning and measures loss of biodiversity at both global and biome/large ecosystem levels" (Steffen et al. 2015: 1259855-5–1259855-6). The metric for assessing this component is the Biodiversity Intactness Index (BII), which

> assesses change in population abundance as a result of human impacts, such as land or resource use, across a wide range of taxa and functional groups at a biome or ecosystem level using per-industrial era abundance as a reference point.
>
> (Steffen et al. 2015: 1259855-6)

The proposed precaution-based boundary for the first component is ten extinctions per million species per year (MSY), which is estimated to be about ten times the background extinction rate. For the second component, the proposed boundary is 90 percent of the pre-industrial BII. Taken together, these proposed boundaries allow for a departure from a human-free Earth system so as to account for past and future human impacts. Although planetary boundary researchers acknowledge significant uncertainty with regard to both indicators (MSY and BII), they put forth an understanding of biosphere or ecological integrity that avoids the incoherence of defining integrity with reference to pristine, wild nature in which humans have no place.

The studies noted in this section strongly imply that a rigorous yet practicable notion of ecological integrity must acknowledge and accept a conditional level of symbiosis between humans and non-human nature in order to be consistent with a mutually enhancing human-Earth relationship. The challenge becomes determining where along the spectrum from the least impacted to the most anthropogenically transformed ecosystems benchmarks for ecological integrity that are practicable in law and governance systems can and should be drawn at different spatial and temporal scales, accounting for humans as an integral ecosystem component. This determination necessarily reflects the aggregate of the individual and collective intentions of human beings; human intention is thus a critical variable for the human prospect. Understanding the role of human intention in successful ecological and eco-cultural restoration and in examples of sustainable governance of common pool resources (CPRs)[1] provides direction for a global approach for instilling a mutually enhancing human-Earth relationship.

Lessons from ecological restoration and governance of common pool resources

The domains of ecological restoration and eco-cultural restoration comprise a rich source for translating broad ideals tied to ecological integrity into practicable approaches at various scales, taking into account historical information regarding the human-Earth relationship. Both ecological and eco-cultural restoration involve a "process of assisting the recovery of an ecosystem that has been degraded, damaged, or destroyed" (SER 2004), taking into account criteria such as ecological integrity, historical fidelity and community engagement (Higgs 2003). Although both domains must account for the dynamics of interactions between humans and non-human nature, eco-cultural restoration places more explicit emphasis on recovery of mutually supportive cultural practices and ecosystem structure and functioning. Both ecological and eco-cultural restoration put into practical perspective questions such as, "Why restore nature? Which natures are possible and acceptable? Which natures can be maintained and sustained? Which restoration project is more important than other pressing environmental and social problems, and deserves allocation of scarce resources?" (Egan et al. 2011: 9). They also inherently emphasize that restoration efforts, and by analogy broader efforts to use law and policy to implement a foundational commitment to notions such as ecological integrity or mutual human-Earth enhancement, always require a communal human choice as to the desired objectives and the means to achieve them.

Human intention is key. Inherent in the proactive domains of ecological and eco-cultural restoration is a human dimension that encapsulates not only the reality of historical human impacts on ecosystems from the local to the global level, but also the deliberate choice that is involved in pursuing "a vision of a better relationship between humans and the rest of the world" (Egan et al. 2011: 1). Because of the deliberate human choice involved, ecological or eco-cultural restoration "is inherently (1) value-laden, (2) context driven, (3) prone to be immersed in disagreement and compromise, and (4) experiential" (Egan et al. 2011: 1–2). The focus on choice distinguishes notions of integrity in ecological or eco-cultural restoration from definitions of ecological integrity that place emphasis on wildness and the absence of anthropogenic interference of any kind in the trajectory of an ecosystem (Westra 2011). Because complete absence from human impacts is impossible, such purist definitions may go too far in preventing ecological integrity from having practical application; and, in the context of law and governance, which are inherently concerned with providing order to collective human regimes of choice, accounting for the role of conscious human choice in applying a concept like ecological integrity is inevitable.

Although restoration ecology at times links ecological integrity to pristine human-free nature (Higgs 2003), ecological integrity in this context more commonly has a broader connotation that accounts for human presence in ecosystems. For example, Higgs describes integrity as "an all-encompassing term for the various features—resilience, elasticity, stress response, and so on—that allow an ecosystem to adjust to environmental change" (Higgs 2003: 122). This notion of integrity inherently includes human-induced environmental change, although with conditions and limits.

Within this broader notion of ecological integrity,

> recognizing and understanding the human dimensions of ecological restoration are critical to the success and longevity of all ecological restoration efforts, especially those undertaken at large scales, on public lands, and/or within urban/suburban settings. These are situations where restoration activities move beyond the vision and control of an individual landowner or small group of like-minded people; these activities are community-based efforts that involve the ideas and concerns of many people.
>
> (Egan et al. 2011: 1)

Indeed, the Society of Ecological Restoration considered, but ultimately rejected, defining ecological integrity so as to include "a critical range of variability in biodiversity, ecological processes and structures, regional and historical context, and sustainable cultural practices" (Higgs 2003: 109). Although this definition was seen as too vague (Higgs 2003), it reflects the importance of accounting for the human dimension in discerning when ecological integrity is attained. A particularly elegant expression of the dialectic this entails is the notion of reciprocal restoration (Geist and Galatowitsch 1999; Egan et al. 2011). Reciprocal restoration is "the mutually reinforcing restoration of land and culture such that the repair of ecosystem services contributes to cultural revitalization and renewal of culture promotes restoration of ecological integrity" (Kimmerer 2011: 258). This formulation resonates strongly with the notion of a mutually enhancing human-Earth relationship, and with the pairing of ecosystem services to humans with human services to ecosystems.

Higgs (2003) accounts for this nuanced interpretation of ecological integrity in advancing an approach to ecological restoration grounded in the integrated concepts of focal practice, ecological integrity, historic fidelity and intention. He captures the conditional space for ecologically acceptable human impacts within the concept of ecological integrity by framing it broadly around these factors, while avoiding the malleable notion of ecological health—which is more detached from alignment with historical conditions and trajectories, and therefore more prone to subjective interpretations (Higgs 2003; Westra 2011; Goldberg et al. 2015). The commitment to re-establishing historical eco-cultural trajectories also distinguishes ecological restoration from ecological engineering or design, in which "novel ecosystems" (if they have not already emerged unintentionally) might be created without reference to historical conditions (Balaguer et al. 2011). Higgs (2003) embraces within the notion of ecological restoration a flexible range of options that include a human presence and that maintains (1) a basis for mutual flourishing of the human and non-human spheres of nature, (2) ties to historical conditions in which a mutually enhancing human-Earth relationship existed, (3) insistence on core notions of resilience and persistence in accordance with broad historical socio-ecological trajectories and (4) recognition of embedded, resilient and communal connections of people and place.

Flexibility in the meaning of ecological integrity can be incorporated into the selection of reference ecosystems that are used to establish objectives for restoration, or more broadly for socio-ecological integrity. Thus, Balaguer et al. (2014) advocate for selection of "*locally-tailored* historical references, using all the available and appropriate conceptual tools, so as to integrate both latent and ongoing ecological and socio cultural processes and values" (Balaguer et al. 2014: 13, emphasis in original). They assert that restoration "should be approached not only with regard to theoretical community ecology and so-called assembly rules theory, but also in the context of an historical sequence and *ecosystem trajectory,* which also involves land use changes, physical processes and landform dynamics" (Balaguer et al. 2014: 13 emphasis in original). However, because ecological restoration is often concerned with re-establishing a damaged ecosystem's historical trajectory as it was prior to human disturbance, the human impacts that may be incorporated into an appropriate historical reference are limited to those that were integral to that pre-existing trajectory (Higgs 2003; Balaguer et al. 2014).

To meet the challenge of identifying an appropriate line between human impacts that may and may not be included in the historical trajectory, Balaguer et al. (2014: 16) assert it is necessary to "retrace the historical scenario of transformation leading the ecosystems and landscapes at issue to their present condition, and then to develop a multiple sequential reference model in three steps." These steps essentially involve (1) identifying the successive states of local ecosystems and landscapes, including those to which human actions contributed, leading to and including the disturbance of concern, (2) establishing a future trajectory that incorporates "the processes of reinforcement, reconnecting, and recuperating lost biodiversity, ecosystem functioning, and the delivery of multiple ecosystem services to multiple beneficiaries" (Balaguer et al. 2014: 17) and (3) discussing the proposed restoration plan with concerned stakeholders.

In his documentary film *Hope in a Changing Climate*, Liu (2009) provides good examples of how landscapes can be transformed in conformance with this kind of approach. The lead example is the ten-year transformation of the Loess Plateau in China from an eroded, over-grazed and poverty-stricken region to a fertile valley with terraced agriculture, greatly improved water management and dedication of higher elevation areas to conservation and water retention. As well, project managers overcame resistance of farmers in the region by briefing them on its goals and methods and involving them in its implementation. The film recounts similar projects in Ethiopia and Rwanda. Other inspiring examples include the ecosystem-transforming reintroduction of wolves in Yellowstone National Park, reforestation efforts through Wangari Matthai's Green Belt Movement (Ciccarese et al. 2012) and the restoration of the channelized Kissimmee River in Florida so as to reinstate its ecological integrity (Whalen et al. 2002).

This discursive and expansive notion of ecological integrity resonates with studies of social systems that have maintained sustainable, enduring use of at least some CPRs in supporting ecosystems (Ostrom 1990). From these studies, Ostrom identified key features of social systems that maintain human provisioning from ecosystems of at least some human needs in perpetuity, without reliance on privatization or government regulation. Those features (Ostrom 1990: 90) are:

- **Clearly defined boundaries.** Individuals or households who have rights to withdraw resource units from the CPR must be clearly defined, as must the boundaries of the CPR itself;
- **Congruence between appropriation and provision rules and local conditions.** Appropriation rules restricting time, place, technology, and/or quantity of resource units are related to local conditions and to provision rules requiring labor, material and/or money;
- **Collective-choice arrangements.** Most individuals affected by the operational rules can participate in modifying the operational rules;
- **Monitoring.** Monitors, who actively audit CPR conditions and appropriator behavior, are accountable to the appropriators or are the appropriators;
- **Graduated sanctions.** Appropriators who violate operational rules are likely to be assessed graduated sanctions (depending on the seriousness and context of the offense) by other appropriators, by officials accountable to these appropriators, or by both;
- **Conflict-resolution mechanisms.** Appropriators and their officials have rapid access to low-cost local arenas to resolve conflicts among appropriators or between appropriators and officials;

- **Minimal recognition of rights to organize.** The rights of appropriators to devise their own institutions are not challenged by external government authorities;
- **Nested enterprises.** Appropriation, provision, monitoring, enforcement, conflict resolution, and governance activities are organized in multiple layers of nested enterprises.

Ostrom emphasized that the emergence of these features, and of the collective constraint on human demands on supportive ecosystems, is contingent on a strong communal attachment to place and the local community, and to intergenerational continuity. These place-based and intergenerational commitments imply a need to adapt legal and governance structures and rules for sustainable and shared human use of supportive ecosystems in response to information from monitoring of relevant social and ecological indicators at a local scale (Ostrom 1990).

A discursive and adaptive approach to giving meaning to ecological integrity, informed by lessons from ecological restoration, paves a potential way to account for difficult challenges with respect to so-called "novel ecosystems" (Seasted et al. 2008; Hobbs et al. 2009; Hobbs et al. 2014; Murcia et al. 2014). The notion of novel ecosystems arose from the inevitable realization that humanity has had an impact on all of Earth's ecosystems, such that some may have "a greater incidence of native and introduced species living under new environmental conditions and new or altered disturbance regimes" (Seasted et al. 2008: 547). Yet, while some novel ecosystems impede a mutually enhancing human-Earth relationship, others tend to favor such a relationship. Thus, coherent criteria are needed to make forward-looking legal and policy decisions along the continuum from pristine to "hybrid" to novel to impaired ecosystems.

A challenge in applying a cohesive yet practicable notion of ecological integrity is that ecosystems at the local, landscape and regional level inevitably and increasingly are subject to human impacts that derive from spatially and temporally diverse drivers—including those that are remote geographically or whose impacts are temporally delayed. Thus, at whatever temporal or spatial scale, practical applications of ecological integrity to promote a mutually enhancing human-Earth relationship require a multi-scalar systems approach that maintains a focus on and connection to the local scale. In Chapter 15, I describe some innovative thinking that has emerged in the degrowth movement on how to integrate autonomy that supports a mutually enhancing human-Earth relationship at the local level with these multi-scale and cross-scale complexities.

Attachment to place and the wisdom of Indigenous legal traditions

Indigenous legal traditions are a rich source for giving meaning and depth to ecological law, in particular, in regard to the importance of attachment to place. Drawing heavily from Indigenous legal traditions in giving definition to ecological law makes sense because a human-inclusive ecocentric approach to law is better in the long term for both humans and non-human nature than that of the failing dominant anthropocentric one. The more enduring moral truths regarding the relationship between humans and the Earth and non-human nature lie within Indigenous worldviews, not in the human-centered traditions of Western law that are so strongly linked to the current ecological crisis.

Indigenous legal traditions around the world are diverse and heterogeneous, but they typically draw heavily on ecological knowledge and experience.

> Aside from sacred sources, Indigenous peoples also find and develop law from observations of the physical world around them. When considering law from this source, it is often necessary to understand how the earth maintains functions that benefit us and all other beings. This approach to legal interpretation attempts to develop rules for regulation and conflict resolution from a study of the world's behaviour.
>
> (Borrows 2010: 28)

This perspective aligns more and more with what contemporary science is revealing about the ecological impacts of human activities in a global community predominantly committed to endless economic growth. Yet, contemporary science is predominantly contained within Western knowledge systems and ways of knowing. Indigenous law typically goes further and deeper, with its grounding in a worldview that developed over centuries of observing, learning from and co-evolving with the ecosystems in which humans were embedded. Thus, Indigenous legal traditions tend to reflect cultural notions of respect, reciprocity and systems relationships (Kimmerer 2013) that are stunted in dominant contemporary law. These notions are at the heart of common Indigenous customs and traditions, such as the honorable harvest (taking respectfully no more than what is truly needed), deep knowledge and appreciation of place, reverence and respect for elders and wise teachers, slow and patient learning in tune with nature's rhythms, and a powerful sense of reciprocity and responsibility to other members of the entire community of life (Brown and Garver 2009, Kimmerer 2013, Borrows 2010). Law that truly integrates these ideas would not treat trees, forests and mountains as persons, but would rather see them as direct representations of the law, of "eco-logics" that should guide humans toward ethical, lawful conduct (Anker 2017).

The emergent governance systems that local communities have developed to maintain sustainable human use of their supporting ecosystems also provide insights into what it might mean to give naturehood to people (Braverman 2018: 135) in ecological law (Ostrom 1990, Agrawal 2001). Underlying all these governance systems is knowledge of and attachment to place that is deep, intergenerational and enduring, as with Indigenous legal traditions. As well, and especially in view of increasing urbanization of human society, ecological law should be supportive of efforts to keep or put people in touch with nature, physically, emotionally and spiritually (Gross 2019).

Concluding thoughts on ecological integrity and attachment to place

Several key conclusions are emerging regarding the foundations of a legal system that seeks to support a mutually enhancing human-Earth relationship.

First, ecological integrity is reflected in many formal and informal instruments in law, from the local to the global. Yet, any practicable notion of ecological integrity or of what is natural must include space for humanity as an integral part of nature and the Earth's ecosystems. Defining ecological integrity with reference only to wild, pristine nature free of human impacts leaves no room at all for humanity if applied from the local to the global level. Discerning if and when the Earth has entered the Anthropocene epoch has brought to light the long history of human alteration of ecosystems, from Paleolithic times to the present. To have meaning across scale, from the local to the global and from the short term to the long term, the concept of ecological integrity must be broad enough to include not only the most pristine areas, but also places where human interventions have significantly and irreversibly changed the human-free evolutionary trajectory, such as novel ecosystems, urban areas and agricultural lands.

Second, because humanity has the capacity to make choices across a broad range of social, political, cultural and economic options, achieving a mutually enhancing human-Earth relationship necessarily requires clear and coherent criteria for controlling individual and collective intentions and choices. In this regard, law plays a key role. So, too, does ecological restoration. Properly done, ecological restoration strictly adheres to a systems-based approach and both commits to and recognizes the need for human intention, choice, decision-making and active intervention to restore ecosystems that human activity has damaged—however damage is defined! Given the long and complicated history of human alteration of Earth's ecosystems, clearly defining what constitutes damage that warrants in situ restoration (as opposed to monetary compensation based on inevitably flawed monetary valuations of ecosystem injuries) is a core challenge for law in the Anthropocene.

Third, to be consistent with an ethic oriented toward achieving and maintaining a mutually enhancing human-Earth relationship, individual and collective human choices must be informed by both contemporary science and traditional ecological knowledge, especially what they say about the behavior of complex adaptive systems. This approach requires the humility that is inherent in dealing with the inevitable uncertainties and unpredictability in how systems evolve by applying a precautionary approach, and not the hubris of many Western secular and religious traditions that idealize total human mastery and perfection of nature. This reconciliation of law with contemporary science and traditional ecological knowledge is at the heart of the transition to law committed to a mutually enhancing human-Earth relationship.

In Part IV, I develop further this systems-based approach for moving from contemporary law to ecological law. The central features of this proposed approach are leverage points for affecting changes in complex systems and a framework for assessing the degree of lock-in or lock-out of systems features that impede or enhance progress toward a mutually enhancing human-Earth relationship. These assessments include relevant metrics and indicators of how material and energy flows cause impacts that imperil a mutually enhancing human-Earth relationship.

Note

1 CPRs are resources (e.g. water, timber) to which access is restricted to a defined community, unlike open access resources, to which access is unrestricted.

References

Agrawal, A. 2001. Common property institutions and sustainable governance of resources. *World Development* 29(10): 1649–1672.

Angermeier, P.L., and J.R. Karr. 1994. Biological integrity versus biological diversity as policy directives. *BioScience* 44: 690–697.

Anker, K. 2017. Law as forest: eco-logic, stories and spirits in indigenous jurisprudence. *Law Text Culture* 21: 191–213.

Balaguer, L., A. Escudero, J.F. Martin-Duque, I. Mola and J. Aronson. 2014. The historical reference in restoration ecology: re-defining a cornerstone concept. *Biological Conservation* 176: 12–20.

Borrows, J. 2010. *Canada's Indigenous Constitution.* Toronto: University of Toronto Press.

Braverman, I. 2018. Law's underdog: a call for more-than-human legalities. *Annual Review of Law and Social Sciences* 14: 127–144.

Bridgewater, P., R.E. Kim and K. Bosselmann. 2014. Ecological integrity: a relevant concept for international environmental law in the Anthropocene? *Yearbook of International Environmental Law* 25(1): 61–78.

Brown, P.G., and G. Garver. 2009. *Right Relationship: Building a Whole Earth Economy.* San Francisco: Berrett-Koehler.

Ciccarese, L., A. Mattsson and D. Pettenella. 2012. Ecosystem services from forest restoration: thinking ahead. *New Forests* 43: 543–560.

De Leo, G.A., and S. Levin. 1997. The multifaceted aspects of ecosystem integrity. *Conservation Ecology* 1(1): 3 [online]: www.consecol.org/vol.1/iss1/art3/

Diamond, J. 2005. *Collapse: How Societies Choose to Fail or Succeed.* New York: Viking Press.

Earth Charter. 2020. Earth Charter website [online]: http://earthcharter.org.

Egan, D., E.E. Hjerpe and J. Abrams, eds. 2011. *Human Dimensions of Ecological Restoration: Integrating Science, Nature, and Culture.* Washington: Island Press.

Ellis, E.C. 2013. Sustaining biodiversity and people in the world's anthropogenic biomes. *Current Opinion in Environmental Sustainability* 5: 368–372.

Ellis, E.C., and N. Ramankutty. 2008. Putting people in the map: anthropogenic biomes of the world. *Frontiers in Ecology and the Environment* 6(8): 439–447.

Foley, S.F., D. Gronenborn, M.O. Andreae, J.W. Kadereit, J. Esper, D. Scholz, U. Pöschl, D.E. Jacob, B.R. Schöne, R. Schreg, A. Vött, D. Jordan, J. Lelieveld, C.G. Weller, K.W. Alt, S. Gaudzinski-Windheuser, K-C. Bruhn, H. Tost, F. Sirocko and P. Crutzen.

2013. The Palaeoanthropocene—the beginnings of anthropogenic environmental change. *Anthropocene* 3: 83–88.

Geist, C., and S.M. Galatowitsch. 1999. Reciprocal model for meeting ecological and human needs in restoration projects. *Conservation Biology* 13(5): 970–979.

Goldberg, M.S., G. Garver and N.E. Mayo. 2015. Revisiting the Metaphor of Human Health for Assessing Ecological Systems and Its Application to Ecological Economics, book chapter in P.G. Brown and P. Timmerman, eds., *Ecological Economics for the Anthropocene*. New York: Columbia University Press.

Gross, M. 2019. Are we losing touch with our world? *Current Biology* 29: R265–R268.

Gunderson, L.H., and C.S. Holling, eds. 2002. *Panarchy: Understanding Transformations in Human and Natural Systems.* Washington, DC: Island Press.

Higgs, E. 2003. *Nature by Design.* Cambridge, MA: The MIT Press.

Hobbs, R.J., E. Higgs, C.M. Hall, P. Bridgewater, F.S. Chapin III, E.C. Ellis, J.J. Ewel, L.M. Hallett, J. Harris, K.B. Hulvey, S.T. Jackson, P.L. Kennedy, C. Kueffer, L. Lach, T.C. Lantz, A.E. Lugo, J. Mascaro, S.D. Murphy, C.R. Nelson, M.P. Perring, D.M. Richardson, T.R. Seastedt, R.J. Standish, B.M. Starzomski, K.N. Suding, P.M. Tognetti, L. Yokob and L. Yung. 2014. Managing the whole landscape: historical, hybrid, and novel ecosystems. *Frontiers in Ecology and the Environment* 12(10): 557–564.

Hobbs, R.J., E. Higgs and J.A. Harris. 2009. Novel ecosystems: implications for conservation and restoration. *Trends in Ecology & Evolution* 24(11): 599–605.

Kim, R.E., and K. Bosselmann. 2013. International environmental law in the Anthropocene: towards a purposive system of multilateral environmental agreements. *Transnational Environmental Law* 2: 285–309.

Kimmerer, R. 2011. Restoration and Reciprocity: The Contributions of Traditional Ecological Knowledge, chapter 18 in D. Egan, E.E. Hjerpe and J. Abrams, eds., *Human Dimensions of Ecological Restoration: Integrating Science, Nature, and Culture.* Washington: Island Press.

Kimmerer, R.W. 2013. *Braiding Sweetgrass.* Minneapolis: Milkwood Editions.

Leopold, A. 1949. *A Sand County Almanac*. New York: Oxford University Press.

Lewis, S.L., and M.A. Maslin. 2015. Defining the Anthropocene. *Nature* 519: 171–180.

Liu, J. 2009. *Hope in a Changing Climate,* documentary film posted on YouTube by Grace Media in 2013 [online]: www.youtube.com/watch?v=bLdNhZ6kAzo.

M'Gonigle, M. 2008. Green legal theory: a new approach to the concept of environmental law. *Neue Konzepte* 4: 34–38.

Mace, G.M., B. Reyers, R. Alkemade, R. Biggs, F.S. Chapin III, S.E. Cornell, S. Diaz, S. Jennings, P. Leadley, P.J. Mumby, A. Purvis, R.J. Scholes, A.W.R. Seddon, M. Solan, W. Steffen and G. Woodward. 2014. Approaches to defining a planetary boundary for biodiversity. *Global Environmental Change* 28: 289–297.

McNeill, J.R. 2003. Observations on the nature and culture of environmental history. *History and Theory* 42(4): 5–43.

Merchant, C. 2003. *Reinventing Eden: The Fate of Nature in Western Culture.* New York: Routledge.

Minteer, B.A. 2012. Geoengineering and ecological ethics in the Anthropocene. *Bioscience* 62(10): 857–858.

Murcia, C., J. Aronson, G.H. Kattan, D. Moreno-Mateos, K. Dixon and D. Simberloff. 2014. A critique of the "novel ecosystem" concept. *Trends in Ecology & Evolution* 29(10): 548–553.

Ordonez, C., and P.N. Duinker. 2012. Ecological integrity in urban forests. *Urban Ecosystems* 15: 863–877.

Ostrom, E. 1990. *Governing the Commons: The Evolution of Institutions for Collective Action.* New York: Cambridge University Press.

Rockström, J., W. Steffen, K. Noone, Å. Persson, F.S. Chapin III, E. Lambin, T.M. Lenton, M. Scheffer, C. Folke, H. Schellnhuber, B. Nykvist, C.A. De Wit, T. Hughes, S. van der Leeuw, H. Rodhe, S. Sörlin, P.K. Snyder, R. Costanza, U. Svedin, M. Falkenmark, L. Karlberg, R.W. Corell, V.J. Fabry, J. Hansen, B. Walker, D. Liverman, K. Richardson, P. Crutzen and J. Foley. 2009. Planetary boundaries: exploring the safe operating space for humanity. *Ecology and Society* 14(2): 32 [online] URL: www.ecologyandsociety.org/vol.14/iss2/art32/

Ruddiman, W.F. 2013. The Anthropocene. *Annual Review of Earth and Planetary Sciences* 41: 45–68.

Ruddiman, R., S. Vavrus, J. Kutzbach and F. He. 2014. Does pre-industrial warming double the anthropogenic total? *The Anthropocene Review* 1(2): 147–153.

Seasted, T.R., R.J. Hobbs and K.N. Suding. 2008. Management of novel ecosystems: are novel approaches required? *Frontiers in Ecology and the Environment* 6(10): 547–553.

SER—Society for Ecological Restoration International Science & Policy Working Group. 2004. *The SER International Primer on Ecological Restoration.* Tucson: Society for Ecological Restoration International. URL www.ser.org.

Steffen, W., K. Richardson, J. Rockström, S.E. Cornell, I. Fetzer, E.M. Bennett, R. Biggs, S.R. Carpenter, W. de Vries, C.A. de Wit, C. Folke, D. Gerten, J. Heinke, G.M. Mace, L.M. Persson, V. Ramanathan, B. Reyers and S. Sörlin. 2015. Planetary boundaries: guiding human development on a changing planet. *Science* 347(6223): 1259885-1–1259885-10.

United Nations. 2012. *The Future We Want.* New York: United Nations [online]: https://documents-dds-ny.un.org/doc/UNDOC/GEN/N11/476/10/PDF/N1147610.pdf?OpenElement.

Westra, L. 2011. Governance for Integrity? A Necessary but Distant Goal, pp. 324–336 in P. Burdon ed., *Exploring Wild Law: The Philosophy of Earth Jurisprudence.* Kent Town, S. Australia: Wakefield Press.

Westra, L., P. Miller, J.R. Karr, W.E. Rees and R.E. Ulanowicz. 2000. Ecological Integrity and the Aims of the Global Integrity Project, pp. 19–41 in D. Pimentel, L. Westra and R.F. Ness, eds., *Ecological Integrity: Integrating Environment, Conservation, and Health.* Washington, DC: Island Press.

Whalen, P.J., L.A. Toth, J.W. Koebel and P.K. Strayer. 2002. Kissimmee River restoration: a case study. *Water Science and Technology* 45(11): 55–62.

Wicklum, D., and R.W. Davies. 1995. Ecosystem health and integrity? *Canadian Journal of Botany* 73: 997–1000.

Part IV

Getting from environmental to ecological law

13 A systems-based guide for moving from environmental to ecological law

In this chapter, I present a flexible, systems-based tool for developing priorities and strategies for moving toward ecological law. The tool is systems-based because it takes into account different kinds of leverage points for initiating changed responses in legal systems, as well as ways in which legal systems resist or are open to change. It is flexible in that different users, such as activists, academics or policymakers, should be able to adapt it to their contexts and needs. My presentation of this lock-in/lock-out assessment tool is framed around three main queries.

First, how can consideration of leverage points for triggering change in systems structure and function help in developing strategies for intervening in legal systems, or systems with which law interacts, so as to orient those systems in desired ways? The response to this inquiry derives primarily from the description and analysis in Meadows (1999) of 12 types of intervention, ranging from minor tinkering with limited impact to transformative changes to the system's overarching paradigm. Transformation of legal systems per se appears to be less important than transformation of how law mediates humanity's relationship with Earth systems with finite life support capacity. From this broader perspective, law and governance regimes act as a key control system that regulates positive and negative feedbacks and other system functions within other systems in the global panarchy (Meadows 1999; Beddoe et al. 2009; Kim and Bosselmann 2013).

Second, how can assessment of the degree of lock-in of features of legal systems or other systems with which the legal system interacts (from features that are completely intractable to those that are more flexible) enhance strategies for fostering and enhancing a mutually enhancing human-Earth relationship? The degree of lock-in of a system feature or relationship is a measure of the system's resistance to change. In assessing the degree of lock-in within a system, the roles of path dependence, resilience and adaptiveness are particularly important. Path dependence, which relates to the importance of a system's initial conditions, limits the range of a system's possible futures, and can both lock in and lock out certain system characteristics. Resilience and adaptiveness are value-neutral system features (that is, they are inherently neither desirable nor undesirable) that play a role in regulating the degree of lock-in of system elements and capacities. Thus, a system's resilience and adaptiveness, and the degree of lock-in of system features, all relate to the system's ability to retain or enhance desirable characteristics and to overcome or resist undesirable ones—where, in a socially constructed system such as law, desirability depends on the goals and objectives tied to an overall paradigm or worldview. Something is locked out of a system if resistance to it is locked in, so lock-in can be thought of as including both lock-in and lock-out.

And third, how can human societies transition toward an enduring, mutually enhancing human-Earth relationship in the Anthropocene using appropriate ecological and other metrics and indicators, along with systems-based tools such as frameworks for assessing leverage points and degrees of lock-in? The promise of such an approach rests on the notion that "[t]o escape a situation of lock-in with multiple, reinforcing maladapted cultural variants, societies can foresee potential decline and develop other cultural variants, thereby allowing a positive regime shift, or one with merely temporary setbacks, thus changing the course of the future" (Beddoe et al. 2009: 2484). The key indicators in this analysis are grounded in the relationship of human manipulation of material and energy stocks and flows with Earth's life support systems. With this framing, the planetary boundaries concept and framework (Rockström et al. 2009; Steffen et al. 2015) and the IPAT framework (Ehrlich and Holdren 1972) are focal, and other indicators, such as ecological footprint and human appropriation of net primary production (HANPP), are also relevant. These indicators are used to demonstrate how assessment of leverage points and degrees of lock-in might be applied to yield strategies and priorities for moving toward and preserving a mutually enhancing human-Earth relationship at and across all scales.

Leverage points for effecting desired changes in law and related systems

Leverage points in a complex system are opportunities to intervene in the system, often in counterintuitive ways, "where a small change in one thing can produce big changes in everything" (Meadows 1999: 1). The effectiveness of different categories of leverage points in producing change in a system varies, depending generally on how deeply embedded they are in the system's structure. Meadows (1999) proposed a hierarchy of 12 categories of leverage points, from the least (parameters and constants) to the most (worldviews and system goals) effective in inducing change in the system. These categories can be grouped into three "transformative layers" within the system (UNEP 2012): (1) a "surface layer" consisting of "feedbacks and adjustments," (2) a "mid layer" consisting of rules and incentives, and (3) a "deep layer" consisting of mindsets or paradigms. More specifically, as explained in the context of sustainability:

> The more embedded drivers are in the system, the more enduring and profound the change needs to be: shifting mindsets is at the core of transformation because it translates into visions, goals and collective action. Changing

> rules and incentives is literally a game changer in the pursuit of transformation, because the right incentives can create structural change and influence key drivers. The outer layer of transformation involves creating and listening to feedback and making adjustments to environmental pressures to maintain progress towards sustainability.
>
> (UNEP 2012: 422)

Generally, the higher (or deeper) the leverage point in Meadows's hierarchy, "the more the system will resist changing it" (Meadows 1999: 19)—meaning change is possible, but the barriers tend to be greater. By contrast, changing parameters (level 12) can produce big changes (or small ones) but rarely changes behavior (Meadows 1999).

Leverage points exist in legal systems just as they do in other complex systems. Table 13.1 gives examples of leverage points in legal systems according to Meadows's hierarchy. Below, an analysis is presented of how Meadows's leverage points framework can help shift legal systems and other complex systems with which the law interacts from the current dominant growth-insistent paradigm toward a mutually enhancing human-Earth relationship.

Assessing lock-in that impedes or enhances system transformation

A key defining characteristic of legal systems and the ecological, social and institutional systems with which they interact is the extent to which they lock in features or behaviors that either impede or improve the prospects for attaining desired objectives. Along with path dependence, resilience and adaptiveness are key systems concepts in this regard, and both are value-neutral, in that they tend to keep in place not only those system structures or functions that advance, but also those that impede progress toward a desired value-laden goal (Ruhl 2011). To assess the role of complex adaptive legal systems in the Anthropocene and in potential responses to the dangerous ecological disruption that marks this new period in history, consideration must extend to the lock-in features of both the legal system and the other complex adaptive systems with which the law interacts—and for which the existence or lack of legal rules might itself be a strong lock-in feature (Ruhl 2011; Ruhl 2012). I present a more detailed analysis of how policy-oriented researchers and decision-makers could apply a lock-in assessment framework to transition the legal system and related systems toward a mutually enhancing human-Earth relationship below.

If a system feature is absolutely locked in or locked out, for example as a consequence of the system's path dependence, then any strategy to alter, add or remove that feature would be futile. Therefore, the strategic framework proposed in this chapter involves assessment of the degree or extent of lock-in. In such an assessment, the resilience and adaptiveness of a system have primary importance; the leverage points discussed above are also relevant. Resilience is a system's capacity "to experience shocks while retaining essentially the same

Table 13.1 Leverage points in legal systems

Leverage point (per Meadows 1999)	*Examples*
12 (parameters)	• Air quality standards • Tax rates • Number of legislators in a legislative body • Number of judges on a court
11 (buffers)	• Offset, bubble and cap-and-trade provisions in US air pollution laws • Sentencing discretion for judges
10 (stock and flow structures)	• Judicial appeals processes • Constitutional procedures for adopting or amending legal provisions • Evolution of law in common law systems
9 (delays)	• Length of judicial processes (e.g. from time crime is committed to when sentence is imposed and served) • Length of time to adopt and implement new laws or regulations
8 (negative feedback)	• Checks and balances among executive, legislative and judicial branches in US constitutional law • Cross examination in a trial • Pollution taxes (Meadows 1999)
7 (dampen positive feedback)	• Evidentiary rules in a trial that limit certain kinds of evidence • Progressive income tax
6 (information flows)	• Mandatory product labeling • Pollutant release and transfer registries (such as the US Toxic Release Inventory (Meadows 1999)) • Freedom of information laws
5 (rules)	• Adoption of a Constitutional structure for national laws • Relationship of national to sub-national law • Rules of procedure in common law and civil law systems
4 (system structure)	• Proposals to create a World Environment Organization or World Environmental Court • Formation of the European Union • Creation of a broad regime of air quality standards and pollution controls • Rights-based frameworks
3 (goals)	• Economic growth • Mutually enhancing human-Earth relationship • Ecological justice
2 (mindset, worldview)	• Anthropocentric • Human-inclusive ecocentric • Wild law (Cullinan 2011)
1 (transcend paradigms)	• Global plurality of diverse paradigms at local, regional, national or international scales

function, structure, feedbacks, and therefore identity" (Ruhl 2011: 1375–1376; Garmestani and Benson 2013). Two general kinds of resilience are relevant here: (1) "engineering resilience," by which the system tends to maintain a specific system state that is highly controlled, predictable, efficient, uniform and constant—for example, in an industrial process in which consistent product quality is desired, and (2) "ecological or ecosystem resilience," by which the system fends off disturbances that could alter its overall structure, function and behavior by developing a broad-based defense that relies predominantly on diversity, adaptiveness, far-from-equilibrium conditions and unpredictability—for example, a tropical forest with high biodiversity (Gunderson and Holling 2002; Hummel et al. 2011; Garmestani and Allen 2014). In non-human systems, ecological resilience, not engineering resilience, prevails. In social-ecological systems, social-ecological resilience, by analogy to ecological resilience, emphasizes learning, innovation and transformability—that is, features that enable the system to adapt and evolve while maintaining its essential structure, functioning and integrity despite disturbances (Hummel et al. 2011). Adaptiveness is the capacity of a system to alter resilience strategies while maintaining its essential structure and function (Ruhl 2011). In complex socially constructed systems that, like law, inevitably interact with ecosystems and other social systems, adaptiveness requires a "focus on maintaining the resilience and adaptive capacity of both social and ecological systems, including such subsystems as institutions and communities" (Garmestani and Allen 2014: 8).

By way of example, the global economic system has features with a significant degree of lock-in—such as a strong commitment to growth, capitalism, strong State sovereignty and property rights and relatively unrestricted markets—that (among other effects) have historically maintained varying degrees of inequality in wealth and income. In turn, unequal income and wealth can lead to unequal access of individuals and organizations to the political process at different scales, and the political system may have its own lock-in tendencies that help maintain those arrangements (Hardoon 2015). Further, the result of this special access may be strong protection of the interests of wealthy individuals or organizations to procure rights and freedoms to engage in activities with detrimental ecological impacts, which may be of global significance, as with climate change. The economic, political and ecological systems at stake interact as well with the legal system, which may contain several features with varying degrees of lock-in that reinforce wealth inequalities and unequal access to the political process. If, for example, constitutional provisions are invoked to maintain such arrangements, as in the case of the United States Supreme Court's decision to reject limits on independent campaign expenditures by individuals and corporations in *Citizens United v. Federal Election Commission*,[1] the degree of lock-in is very strong, given the difficulty of amending the United States Constitution or overturning Supreme Court precedents. Such resistance to change has impacts not only in the legal system, but also in the social and ecological systems with which it interacts.

Using lock-in assessment, leverage points and relevant metrics and indicators to transition to a legal system that supports mutual human-Earth enhancement

A core challenge for law in the Anthropocene is to assess and respond to lock-in features that affect the potential of the legal system, as well as of the social, ecological, technological and institutional systems with which it interacts, to promote a mutually enhancing human-Earth relationship (Beddoe et al. 2009). Such an assessment should include three main categories of lock-in features of those systems: (1) existing features that promote a mutually enhancing human-Earth relationship, (2) existing features that undermine a mutually enhancing human-Earth relationship and (3) new features that would advance a mutually enhancing human-Earth relationship. Each category can be further subdivided according to the spatial or institutional scale (from local to global) at which the lock-in operates. For each category, the assessment at different scales should account for (1) the degree of lock-in, (2) the nature and importance of the ecological and social stakes the lock-in affects and (3) strategies and reasonable time-frames for responding—by removing or preventing undesirable lock-in and by enhancing or integrating lock-in that is favorable.

These assessments should take a systems approach, with ongoing consideration of how various features with lock-in tendencies interrelate across systems and across temporal and spatial scales. In most cases, a degree of lock-in that affects one complex adaptive system, like law, is almost certain to affect other social and ecological systems, as well as integrated social-ecological systems and perhaps the full global panarchy—as with the example of the *Citizens United* case. Taken together, these assessments and the leverage points discussed previously provide options for developing strategies and priorities for moving toward a mutually enhancing human-Earth relationship. Because of the uncertainties and complexities inherent in applying leverage points, "a policy approach that diversifies strategy across all the layers of transformation presents a diverse and resilient portfolio for using leverage to its greatest advantage" (UNEP 2012: 422).

Adaptiveness and precaution are key features of a legal system supportive of a mutually enhancing human-Earth relationship (Garver 2013) and should be incorporated into responses based on assessments of lock-in and potential leverage points. Adaptive approaches require more robust systems monitoring and response than exist in typical environmental and natural resources law regimes (Ruhl 2008; Craig 2013; Garver 2013). Historic or other types of systems-derived baselines or references play an important role in adaptive approaches (McNeill 2003). A key challenge is to identify times and places when the human-Earth relationship appeared to be in a desirable, mutually enhancing systems trajectory (Ruhl and Salzman 2011).

In some instances, relatively static baselines may be useful, at least temporarily, as in the case of referring to greenhouse gas emissions in 1990 as a historic baseline for climate change mitigation (Ruhl and Salzman 2011). However, the "stationarity" that is assumed in many aspects of environmental law, and that might be associated with historic baselines, is problematic given the dynamic and non-linear evolution of ecosystems (Craig 2013). Because there is no going back in time, the lessons to take from historic examples should emphasize the systems dynamics involved in succeeding or failing to fend off catastrophic change—the role or absence of positive and negative feedback loops and the time lags involved in attempting to modify or enhance them, for example (Anderies and Janssen 2013). Further, the exact nature and functioning of an emergent social-ecological system, or a global panarchy, that is consistent with a mutually enhancing human-Earth relationship is inherently unknowable, given the surprise and irreducible uncertainty that are essential characteristics of complex adaptive systems. Thus, the systems goals for policy development should be more broadly based on reference conditions that incorporate the evolutionary capacity of systems to adapt and maintain resilience. For example, as explained in detail in Chapter 12, ecological restoration involves intentional intervention to return to a desired trajectory based on objectives that integrate historical information—both socio-cultural and ecological—and criteria of ecological integrity (Higgs 2003).

Metrics and indicators of the extent to which the human-Earth relationship is mutually enhancing should also be incorporated into the development of lock-in assessments and response strategies they support. Planetary boundaries, which are being developed and refined with a systems-based methodology (Rockström et al. 2009; Steffen et al. 2015), are treated here as the main framework for relevant metrics and indicators because they reflect the impact of human manipulation of material energy stocks and flows on Earth's life support systems. They and other relevant metrics and indicators were presented in Chapter 8.

A comprehensive set of assessments as outlined in this section would enhance the development of coordinated adaptive strategies to prioritize responses to lock-in challenges according to their ecological or social stakes and the ease with which the lock-in challenge can be addressed at different scales.

Assessing pathways toward a mutually enhancing human-Earth relationship: framework for lock-in assessments using leverage points and relevant indicators in the case of climate change

The lock-in assessment approach outlined in this chapter may be applied in the context of the human community's response to potentially catastrophic climate change. Although the climate change boundary of 350 ppm of carbon dioxide in the atmosphere has already been exceeded (Rockström et al. 2009; Steffen et al. 2015), humanity has not yet surpassed the estimated outer limit of the zone of uncertainty for climate change of 450 ppm (Steffen et al. 2015). Here I present an outline for key aspects of a lock-in assessment based on a possible pathway for addressing climate change in a manner consistent with a mutually enhancing human-Earth relationship.

"The planetary boundaries framework arises from the scientific evidence that Earth is a single, complex, integrated system—that is, the boundaries operate as an interdependent set" (Steffen et al. 2015: 1259855–8). Thus, a systems-based lock-in assessment, combined with systems-based leverage points for effecting change in the human-Earth system, is well-suited to the planetary boundaries—although assessment of the degree of lock-in with respect to any individual boundary, such as climate change, must not lose sight of the interdependence of all the boundaries.

The pathway used here for assessing lock-in related to climate change is based on reaching a reliable global commitment to return the concentration of carbon dioxide in the atmosphere to 350 ppm (Steffen et al. 2015). Although this level of CO_2 has been considered necessary to prevent average global temperatures from rising more than 2 °C above the pre-industrial average (typically, using a base period of 1880 to 1920 (Hansen et al. 2013)), some prominent climate researchers suggest that even a 2 °C rise would have catastrophic consequences, given the severity of climate change impacts (e.g. intensity, frequency and duration of heat waves and other severe weather events; melting of glaciers and polar ice sheets) associated with the global warming of over 0.8 °C that already has occurred (Hansen et al. 2013; Steffen et al. 2015).

Returning to 350 ppm CO_2 in the atmosphere within a specified time period can be converted into total cumulative emissions of CO_2. A conservative estimate of total cumulative emissions of CO_2 allowable in order to return to 350 ppm CO_2 in the atmosphere by 2100 is 500 GtC (Hansen et al. 2013).[2] Because cumulative emissions as of around 2010 were approximately 370 GtC, the pathway needed to return to 350 ppm by 2100 requires an aggressive shared international responsibility to determine the source of the remaining 130 GtC for the period from 2010 to 2100. The total allowable cumulative emissions of 500 GtC can be plotted as an emissions curve (emissions plotted against time) of emissions that first rise and then decline to zero, with the area under curve the 500 GtC of total emissions (Hansen et al. 2008, using 500 GtC; Schlesinger et al. 2012, using 1,000 GtC). In either case, the legal and governance challenge is to distribute the emissions reduction curve internationally; that is, to determine what national emissions sources from different countries contribute to the total emissions at any point along the curve. One proposal for fairly distributing the legal obligation to limit emissions internationally is to divide the total cumulative emissions in half, such that by the end of the hydrocarbon phase-out period, less developed countries (non-Annex B countries in the Kyoto Protocol) will have emitted the same amount of CO_2 as developed countries (Annex B countries in the Kyoto Protocol) (Schlesinger et al. 2012). Another option for fair sharing of this obligation is to base the emissions reduction curves on a goal of equal per capita emissions through time.

The pathway I propose would involve ranking all potential hydrocarbon sources globally according to their EROEI and then imposing a binding obligation to strictly favor the sources with the highest EROEI through time. That is, a legal mechanism and supporting governance structure would be necessary to ensure that decisions regarding which hydrocarbon sources to exploit would be based strictly on EROEI, and not on market logic or geopolitics. This pathway is designed to allow the greatest amount of net energy from extracted hydrocarbons while staying within the planetary boundary for climate change, in this case by using the 500 GtC cumulative emissions curve to establish binding emissions limits. Of course, this means that a significant amount of known or discoverable hydrocarbon fuel reserves could not be developed (McGlade and Ekins 2015).

Without question, this pathway to sufficiently regulating climate change would face considerable challenges, both in overcoming obstacles and in implementing new opportunities. Tables 13.2 and 13.3 provide examples of key potential obstacles and opportunities related to this pathway, the systems leverage points that would potentially need to be invoked to overcome or pursue them, the nature and importance of the social and ecological stakes to which they relate, a qualitative estimate of the degree of lock-in (low, medium or high) and a qualitative assessment of the time needed to overcome or pursue them (short, medium or long). Very roughly, "short" is about ten years or less, "medium" is about 10 to 50 years and "long" is more than 50 years.

Conclusions

The proposed prescriptions for law that have emerged from complexity theory scholars are in line in many ways with features that have been identified for "the rule of ecological law" (Garver 2013), another way of framing law for the Anthropocene. In addition to adaptiveness, enhanced monitoring and precaution about transgressing ecological limits, those features include treating humanity as embedded in Earth's ecosystems; giving primacy to ecological limits; fully integrating the rule of ecological law into all areas of law and policy; radically reducing humanity's use of material and energy; establishing global and supranational rules distributed according to the principle of subsidiarity and committing to intragenerational, intergenerational and interspecies fairness (Garver 2013). However, much of the scholarship on law as a complex adaptive system focuses solely or too extensively on environmental law as the domain in which solutions based on adaptive management and precaution should be developed. This focus is too limited. A transformative approach that brings law in line with promoting ecological integrity and a mutually enhancing human-Earth relationship must also fully integrate the legal infrastructure that supports practices in finance, trade, property transactions, contracts, commercial enterprises, taxation and so on, all of which play important roles in the human-Earth relationship.

In the dominant growth-insistent paradigm, technological solutions to environmental challenges are emphasized. The focus for innovation is on technology that will decouple economic growth from negative ecological and social impacts. The role and goal of law is to maintain the reliance of the economic, social and political systems on economic growth and technological innovation (Garmestani and Allen 2014; Capra and Mattei 2015). This overreliance on technological solutions and optimism carries the risk that human society will turn ultimately, and without adequate systems-based understanding and precaution, to untried and potentially catastrophic solutions such as several

Table 13.2 Lock-in assessment: obstacles to coherent climate change law and policy

Examples of obstacles	*Key leverage points (per Meadows 1999)*	*Nature and importance of social and ecological stakes*	*Degree of lock-in*	*Time to overcome*
Growth-insistence	1, 2, 3, 4, 5	In a growth-insistent economy, maintaining a growing economy has primacy over ecological limits. Resource productivity that allows perpetual growth is ultimately insufficient to limit greenhouse gas (GHG) emissions enough to meet planetary boundaries.	High	Long
Investments (and expected returns) in activities that drive climate change (e.g. unconventional fossil fuel extraction)	2, 3, 5, 6	In the current dominant global economic and finance system, investors expect profit. Overinvestment in extractive activities that drive climate change means that investments will be lost if the planetary boundary for climate change is strictly enforced.	Low-High	Short-Long
Private property rights (US)	1, 2, 5	Strong private property rights regimes make it difficult to adopt new limitations on GHG emissions. In the US, some new constraints might require constitutional amendments.	High	Medium-Long
Global trade regime (WTO, NAFTA, etc.)	2, 3, 4, 5, 6, 7	The global trade regime is a key component of the global commitment to economic growth, based on conventional economics that inadequately considers systems-based consideration of the ecological impacts of material and energy extraction and flows.	Medium-High	Medium
Weak supranational law	2, 3, 4, 5	Strong supranational rules in international law exist only at the regional level with respect to the environment (e.g. the EU) and at the global level only with respect to trade and investment. Yet, they are essential for staying within planetary boundaries	Medium-High	Long
Weak building codes	5, 7, 8, 9, 10, 12	In North America, building codes are starting to impose higher standards for renewable energy, energy efficiency, etc. The building sector is significant in regard to climate change and other ecological impacts, but changes to building codes mostly touch the more superficial transformative layers in terms of leverage.	Low-Medium	Short-Medium

Notes

Meadows's leverage points are coded as follows, per Meadows 1999; Figure 4.2: (1) Transcend paradigms, (2) Mindset, (3) Goals, (4) System structure, (5) Rules, (6) Information flows, (7) Dampen positive feedbacks, (8) Negative feedbacks, (9) Delays, (10) Stock and flow structures, (11) Buffers, (12) Parameters.

Table 13.3 Lock-in assessment: opportunities for coherent climate change law and policy

Opportunities (examples)	*Key leverage points (per Meadows 1999)*	*Nature and importance of social and ecological stakes*	*Degree of lock-in*	*Time to implement*
Enforceable global GHG cap	2, 3, 4, 5, 7, 10, 12	Adoption and enforcement of a global GHG cap, such as a cap on cumulative emissions and a detailed plan for distributing the right to emit, would mark a significant turning point in addressing climate change.	High	Long
Expansion of common/public property	1, 2, 3, 4, 5	Expanding common and public property would likely make it easier to impose restrictions or obligations that would mitigate climate change, both in terms of emissions and deforestation.	Medium-High	Medium-Long
Rational use of planetary boundaries, EROEI and related metrics	2, 3, 4, 7	A strong international commitment to giving the planetary boundaries primacy in law and governance would potentially significantly transform global governance. Commitment to extracting hydrocarbons according to EROEI or similar metrics could be consistent with market-based approaches because it is economically efficient but doing this without a cap on emissions would have limited effect.	Low-High	Short-Long
Global carbon price	4, 9	A global carbon price would likely help transition away from hydrocarbon energy, but tying them to strict planetary boundaries could be elusive, especially if there is no GHG cap.	Low-Medium	Short-Medium
Development of renewable energy alternatives	4, 5, 7, 9, 10	Renewable energy is expanding and will likely continue to expand with little or no change in law or policy. However, energy transitions take a very long time, and subsidies, a lack of full-cost accounting, utilitarianism and other aspects of the dominant economic system, plus the shift in infrastructure needed, impose significant obstacles.	Low-High	Short-Long
Strict green building codes	4, 5, 7, 8, 9	Changes in building codes are typically incremental and relatively infrequent and based on input from a wide array of stakeholders. This can impede adoption of strict codes that are consistent with the planetary boundary for climate change. As well, the building industry is tied to other industries through supply chains that are not easily changed, building turnover rates are relatively slow, and the supporting infrastructure likely has a high degree of lock-in (e.g. urban design and lay-out, transport systems, etc.)	Low-Medium	Short-Medium

Notes

Meadows's leverage points are coded as follows, per Meadows 1999; Figure 4.2: (1) Transcend paradigms, (2) Mindset, (3) Goals, (4) System structure, (5) Rules, (6) Information flows, (7) Dampen positive feedbacks, (8) Negative feedbacks, (9) Delays, (10) Stock and flow structures, (11) Buffers, (12) Parameters.

geoengineering (or climate engineering) proposals (Robock 2008; Kim and Bosselmann 2013; Scott 2013).

With societal goals shifted to fostering socio-ecological integrity and a mutually enhancing human-Earth relationship, the focus for innovation broadens to include innovations in normative regimes such as ethics, economics and law. Although technology that reduces material and energy throughput in the economy and preserves or enhances socio-ecological integrity remains important, much more emphasis is placed on developing alternatives that challenge or reject the orthodoxies of human constructs like economics and law that underlie the growth-insistent paradigm. A systems-based assessment of the potential of existing or innovative features of the legal systems and others with which it interacts to lock in systems behaviors that work to advance or impede ecological integrity and a mutually enhancing human-Earth relationship would help illuminate a more hopeful pathway for law in the Anthropocene. Chapter 5 goes into further detail regarding how the legal contours of current international trade regimes must change to accord more coherently with a transition to a mutually enhancing human-Earth relationship.

Notes

1 558 U.S. 310 (2010).
2 A less conservative estimate is 1,000 GtC of cumulative emissions (Schlesinger et al. 2012). Precaution favors using 500 GtC.

References

Anderies, J.M., and M.A. Janssen. 2013. Robustness of social-ecological systems: implications for public policy. *The Policy Studies Journal* 41(3): 513–536.

Beddoe, R., R. Costanza, J. Farley, E. Garza, J. Kent, I. Kubiszewski, L. Martinez, T. McCowen, K. Murphy, N. Myers, Z. Ogden, K. Stapleton and J. Woodward. 2009. Overcoming systemic roadblocks to sustainability: the evolutionary redesign of worldviews, institutions, and technologies. *Proceedings of the National Academy of Science* 106(8): 2483–2489.

Capra, F., and U. Mattei. 2015. *The Ecology of Law: Toward a Legal System in Tune with Nature and Community*. San Francisco: Berrett-Koehler.

Craig, R.K. 2013. Learning to think about complex environmental systems in environmental and natural resources law and legal scholarship: a twenty-year retrospective. *Fordham Environmental Law Review* 24: 87–102.

Cullinan, C. 2011. *Wild Law: A Manifesto for Earth Justice (2nd ed.)*. White River Junction VT: Chelsea Green.

Ehrlich, P.R., and J.P. Holdren. 1972 (May). Critique of *The Closing Circle*. *Bulletin of the Atomic Scientists* 16–27.

Garmestani, A.S., and C.R. Allen, eds. 2014. *Social-Ecological Resilience and the Law*. New York: Columbia University Press.

Garmestani, A.S., and M.H. Benson. 2013. A framework for resilience-based governance of social-ecological systems. *Ecology and Society* 18(1) art 9 [online]: www.ecologyandsociety.org/vol.18/iss1/art9/.

Garver, G. 2013. The rule of ecological law: the legal complement to degrowth economics. *Sustainability* 5(1): 316–337 [online]: www.mdpi.com/2071-1050/5/1/316/htm
Gunderson, L.H., and C.S. Holling, eds. 2002. *Panarchy: Understanding Transformations in Human and Natural Systems.* Washington, DC: Island Press.
Hansen, J., P. Kharecha, M. Sato, V. Masson-Delmotte, F. Ackerman, D.J. Beerling, P.J. Hearty, O. Hoegh-Guldberg, S. Hsu, C. Parmesan, J. Rockström, E.J. Rohling, J. Sachs, P. Smith, K. Steffen, L. Van Susteren, K. von Schuckmann and J.C. Zachos. 2013. Assessing "dangerous climate change:" required reduction of carbon emissions to protect young people, future generations and nature. *PLOS ONE* 8(12): e81648.
Hansen, J., M. Sato, P. Kharecha, D. Beerling, R. Berner, V. Masson-Delmotte, M. Pagani, M. Raymo, D.L. Royer and J.C. Zachos. 2008. Target atmospheric CO2: where should humanity aim? *Open Atmospheric Science Journal* 2: 217–231.
Hardoon, D. 2015. *Wealth: Having It All and Wanting More*. Oxfam Issue Briefing, January 2015. Oxford UK: Oxfam International.
Higgs, E. 2003. *Nature by Design*. Cambridge, MA: The MIT Press.
Hummel, D., T. Jahn and E. Schramm. 2011. *Social-Ecological Analysis of Climate Induced Changes in Biodiversity—Outline of a Research Concept*. Knowledge Flow Paper No. 11. Frankfurt: LOEWE Biodiversität und Klima.
Kim, R.E., and K. Bosselmann. 2013. International environmental law in the Anthropocene: towards a purposive system of multilateral environmental agreements. *Transnational Environmental Law* 2: 285–309.
McGlade, C., and P. Ekins. 2015. The geographical distribution of fossil fuels unused when limiting global warming to 2°C. *Nature* 517: 187–193.
McNeill, J.R. 2003. Observations on the nature and culture of environmental history. *History and Theory* 42(4): 5–43.
Meadows, D. 1999. *Leverage Points: Places to Intervene in a System.* Hartland VT: The Sustainability Institute.
Robock, A. 2008. 20 reasons why geoengineering may be a bad idea. *Bulletin of the Atomic Scientists* 64(2): 14–18.
Rockström, J., W. Steffen, K. Noone, Å. Persson, F.S. Chapin III, E. Lambin, T.M. Lenton, M. Scheffer, C. Folke, H. Schellnhuber, B. Nykvist, C.A. De Wit, T. Hughes, S. van der Leeuw, H. Rodhe, S. Sörlin, P.K. Snyder, R. Costanza, U. Svedin, M. Falkenmark, L. Karlberg, R.W. Corell, V.J. Fabry, J. Hansen, B. Walker, D. Liverman, K. Richardson, P. Crutzen and J. Foley. 2009. Planetary boundaries: exploring the safe operating space for humanity. *Ecology and Society* 14(2): 32 [online] URL: www.ecologyandsociety.org/vol.14/iss2/art32/
Ruhl, J.B. 2008. Law's complexity: a primer. *Georgia State University Law Review* 24(4): 885–911.
Ruhl, J.B. 2011. General design principles for resilience and adaptive capacity in legal systems—with applications to climate change adaptation. *North Carolina Law Review* 89: 1373–1403.
Ruhl, J.B. 2012. Panarchy and the law. *Ecology and Society* 17(3); art 31 [online]: http://dx.doi.org/10.5751/ES-05109-170331.
Ruhl, J.B., and J. Salzman. 2011. Gaming the past: the theory and practice of historic baselines in the administrative state. *Vanderbilt Law Review* 64(1): 1–57.
Scott, K.N. 2013. International law in the Anthropocene: responding to the geoengineering challenge. *Michigan Journal of International Law* 34: 309–358.
Schlesinger, M.E., M.J. Ring and E.F. Cross. 2012. A revised fair plan to safeguard Earth's climate. *Journal of Environmental Protection* 3: 1330–1335.

Steffen, W., K. Richardson, J. Rockström, S.E. Cornell, I. Fetzer, E.M. Bennett, R. Biggs, S.R. Carpenter, W. de Vries, C.A. de Wit, C. Folke, D. Gerten, J. Heinke, G.M. Mace, L.M. Persson, V. Ramanathan, B. Reyers and S. Sörlin. 2015. Planetary boundaries: guiding human development on a changing planet. *Science* 347(6223): 1259885-1–1259885-10.

United Nations Environment Programme (UNEP). 2012. *GEO5: Global Environment Outlook 5—Environment For the Future We Want*. Nairobi: UNEP.

14 Overcoming ecological challenges of international trade

An illustration

People have traded with each other over increasingly long distances for thousands of years. The effects of this long history of trade on humanity and the human-Earth relationship are momentous. Trade has been a source of enormous wealth and opportunity for many as it spread market goods, services and ideas around the world (Young et al. 2006). It has also facilitated environmental change on a global scale. For example, one proposal for the formal geological beginning of the Anthropocene is the global change in pollen distribution that resulted from the cross-introduction of new plant and animal species following Columbus's voyage to the Americas in 1492 (Lewis and Maslin 2015). The Columbian Exchange coincided with the onset of the fully global trade system from which the contemporary international trade regime emerged. In the modern interconnected world, trade has many benefits; I do not propose an end to international trade. However, much of the environmental change to which trade has contributed has been harmful to the human-Earth relationship, especially in recent times. Trade has led to the introduction of species and diseases in places where they were unknown. As a catalyst of flows of material and energy, it has contributed to pressures on all the Earth system processes that the planetary boundaries represent.

Globalized trade and finance are cornerstones of the dominant growth-insistent economic system in the contemporary world (Beddoe et al. 2009; Orr 2011). Yet, just as the dominant global economic paradigm lacks a coherent way for understanding and responding to the ecological impacts of economic activity, the international community has not adequately reckoned with the significant impacts on Earth's ecosystems that international trade and finance exacerbate (Rees 2002). Consistent with the failings of most national and subnational systems of environmental law, the international approach to the ecological consequences of trade has focused on discrete ecological impacts rather than on a comprehensive, systems-based and holistic look at aggregated and cross-scale effects. This prevailing reductionist approach includes the thermodynamically flawed assumption that economic growth will yield technological innovations that will sufficiently increase the efficiency of material and energy throughput or otherwise address ecological concerns (Sachs 2008: 29–30; UNEP 2011). Moreover, long-distance trade and international investment tend to increase the detrimental influence of remote owners and decision-makers on social, economic and ecological systems at a local scale (Pearce 2012). The current global network of trade is a hard-wired subsystem of the global panarchy that has strong links with and support from legal systems from local to global scales—especially with respect to the strong focus in the law on protection of private property rights and of State sovereignty, and on limited regulation of ever-expanding market regimes.

In this chapter, I apply the concepts and proposals developed in the preceding chapters to the United States' approach to international trade, which is grounded in and reinforces the global growth-insistent paradigm. The United States is used as a lens on the global system of trade and finance because of the significant role of the United States in international trade and finance. The approach to international trade and finance in arenas in which the United States has considerable influence—especially the WTO, the NAFTA and the NAAEC and post-NAFTA trade agreements such as the USMCA—increases the degree of lock-in of system features that impede progress toward a mutually enhancing human-Earth relationship. In particular, analysis of the first 20 years of the NAFTA environmental regime, which is the basic model for the United States' approach to trade and environment in subsequent trade deals, shows that it is a near ideal example of a tragic institution: a form of lock-in tendency by which an institution created to address environmental issues has become ineffective but difficult to change (Daniels 2007). The weak environmental safeguards in the NAFTA model, even with modifications in its replacement, the USMCA, and other subsequent trade agreements, are insufficient to regulate the drivers of ecological pressures associated with increasing international exchange of goods and services and strong protection of foreign investments. Worse, even those weak provisions have not been adequately implemented.

I propose a shift to a human-inclusive ecocentric approach to international trade, built on the goals, concepts and metrics elaborated in this book. This approach would require a global framework founded on an international commitment to a mutually enhancing human-Earth relationship and fair sharing of Earth's life support capacity among present and future generations of humans and the rest of Earth's commonwealth of life (Brown and Garver 2009). Priority in legal systems would be given to place-based decision-making at the local level, nested within a cross-scale system that would integrate higher-scale measures needed to regulate ecological impacts as appropriate (M'Gonigle 2008). Trade and finance in such a framework would shift towards supplementing local needs with the lowest practicable pressure on planetary boundaries and other relevant indicators of the aggregate scale of ecological conditions that bound the human-Earth relationship. However, analysis of the leverage points to pursue and the lock-in tendencies to either overcome or instill indicates that a transition toward such a trade regime will be very challenging.

Historical trajectory of trade

Around 4,000 years ago, domestication of animals capable of transporting heavy loads over long distances, especially camels, led to the development of extensive networks in Asia for trading salt, spices, silk, gems and other valued goods (Ellis et al. 2013). Regional trade networks gradually expanded in similar patterns in other parts of the world (Worster 1987). In Europe, for example, regional trade of grains, fur, stockfish and other goods in the twelfth and thirteenth centuries solidified into the Hanseatic League. The Hanseatic League was a sophisticated network for facilitating trade and protecting the interests of merchants in Northern Europe from the thirteenth to the seventeenth century (Hoffman 2001).

The long-distance oceanic voyages of Europeans to East Asia, Africa and the Americas beginning in the late fifteenth century initiated a period of significant expansion in trade. This period marks the origins of a global trade regime, in which increasingly wealthy population centers in Europe and eventually elsewhere became more and more dependent on provisioning from remote parts of the world—the metabolic rift noted in Chapter 2 (Hoffman 2001; Pomeranz 2002; Jones 2003; Clark and Foster 2009). People, especially in wealthy industrial societies, have become increasingly detached from the ecosystems that maintain them ever since; based on consumption, individual ecological footprints globally are less and less local (Nyquist et al. 2013). As noted in Chapter 2, the world market established by the end of the early modern era, at the dawn of the Industrial Revolution, created an unprecedented level of global interconnectedness and resulting impacts. Powered by fossil fuels and the population expansion, technological innovations and rising consumption since the dawn of the Industrial Revolution, this global market has expanded exponentially ever since (Polanyi 1957). Today, the WTO is the international institution that administers and enforces international trade rules at the global scale, with bilateral or multilateral trade agreements, such as NAFTA and its replacement, the USMCA, supplementing those rules at the regional scale.

Contemporary international trade: the WTO, NAFTA and the USMCA

The contemporary rationale for international trade is embedded in the dominant paradigm that gives priority to economic growth and strong protection of private property and State sovereignty. Trade broadens the market arenas in which private proprietors can seek profit, create employment opportunities and contribute to economic growth. It also expands the goods and services available to people worldwide—for example, North American grocery stores now have produce from around the world throughout the year. Trade liberalization reduces or eliminates measures of sovereign states that impose tariffs or other restrictions on imports or foreign investments.

The main economic justification for international trade derives from Ricardo's principle of comparative advantage—that is, a country "has a comparative advantage if it can produce the good in question more cheaply relative to other goods it produces than can its trading partner, regardless of absolute costs" (Daly and Farley 2004: 310). Comparative advantage relies on a country's internal cost ratios in producing goods and services rather than on absolute advantage;

absolute advantage is based on a direct comparison of costs of individual goods and services in different countries. However, comparative advantage is based on the assumption that capital is immobile (Daly and Farley 2004). In the globalized economy, in which capital is increasingly mobile beyond national borders through foreign investments and transnational companies, this core assumption is more and more in doubt. In contemporary trade regimes,

> [g]lobalization creates an increasingly prominent role for transnational corporations, encourages the transportation of resources and manufactured goods all over the planet, facilitates the instantaneous opportunistic movement of finance capital across national boundaries in search of the highest returns, and generally encourages the integration of regional and national economies.
>
> (Rees 2002: 257)

In this economic environment, the competition for high returns, combined with the domestic benefits nation states derive if they can attract investment and productive enterprises, creates a disincentive for strong environmental protection or other socially or ecologically beneficial measures that reduce profits (Rees 2002). This dynamic is especially disadvantageous for less developed countries, which end up trapped in debt and hampered in regard to domestic social and environmental programs as they compete to export commodities to markets in wealthier countries (Rees 2002).

The international community has a strong commitment to enhancing trade and investment across national borders. Expanded liberalized trade and investment is a key component of the international community's dominant commitment to ever-rising economic growth (Gonzalez de Molina and Toledo 2014). For example, in 2014 the G20 leaders stated:

> Trade and competition are powerful drivers of growth, increased living standards and job creation. In today's world we don't just trade final products. We work together to make things by importing and exporting components and services. We need policies that take full advantage of global value chains and encourage greater participation and value addition by developing countries. Our growth strategies include reforms to facilitate trade by lowering costs, streamlining customs procedures, reducing regulatory burdens and strengthening trade-enabling services. We are promoting competition, entrepreneurship and innovation, including by lowering barriers to new business entrants and investment. We reaffirm our longstanding standstill and rollback commitments to resist protectionism. … We need a strong trading system in an open global economy to drive growth and generate jobs. To help business make best use of trade agreements, we will work to ensure our bilateral, regional and plurilateral agreements complement one another, are transparent and contribute to a stronger multilateral trading system under World Trade Organization (WTO) rules. These rules remain the backbone

> of the global trading system that has delivered economic prosperity. A robust and effective WTO that responds to current and future challenges is essential.
>
> (G20 2014, paras. 8, 16)

The outcome document of the Rio+20 conference stated that measures to promote a green economy or sustainable development should "not constitute a means of arbitrary or unjustifiable discrimination or a disguised restriction on international trade" (United Nations 2012, para. 58). The Sustainable Development Goals (SDGs) that the United Nations General Assembly adopted in 2015 implicitly endorse the WTO regime (United Nations 2015a, SDG 17.10) and include a goal of doubling the share of the least developing countries' global exports by 2020 (United Nations 2015a, SDG 17.11).

The WTO, established in 1995, aims "to help trade flow as freely as possible—so long as there are no undesirable side effects—because this is important for economic development and well-being" (WTO 2020). At the heart of the WTO

> are the WTO agreements, negotiated and signed by the bulk of the world's trading nations. These documents provide the legal ground rules for international commerce. They are essentially contracts, binding governments to keep their trade policies within agreed limits. Although negotiated and signed by governments, the goal is to help producers of goods and services, exporters, and importers conduct their business, while allowing governments to meet social and environmental objectives.
>
> (WTO 2020)

The WTO administers and enforces trade rules pursuant to the General Agreement on Tariffs and Trade (GATT), the General Agreement on Trade in Services (GATS) and the Trade-Related Aspects of Intellectual Property Rights (TRIPS). A significant change from prior international trade arrangements was the establishment of an overarching dispute settlement process covering all aspects of the WTO.

Others have done comprehensive reviews of the WTO agreements and their relationship to environmental issues (Watson 2013). Such a detailed analysis is beyond the scope of this book, but some summary observations regarding the WTO are relevant to consideration of how international trade and investment can best promote a mutually enhancing human-Earth relationship in the Anthropocene. First, the WTO institutionalizes a global economic model that promotes global expansion of economic activity founded on limited interference with private property rights, with no mechanism to understand, monitor or make adjustments according to aggregate impacts on ecological limits, such as planetary boundaries. Second, the dispute settlement mechanism in the WTO makes clear the international community's capacity to adopt supranational rules with an enforcement regime that has real impact on national policy. The final decisions in WTO disputes in which national environmental measures were held to violate WTO rules demonstrate the power of this judicialization (Watson 2013) of international rules and show how difficult it is for WTO Member States to adopt environmental measures that overcome concerns regarding over-regulation and protectionism. The most notable of these cases were the Tuna-Dolphin case (rejecting US import restrictions on tuna caught by methods that harmed dolphins), the Shrimp-Turtle case (rejecting US import restrictions on shrimp caught using methods that harmed sea turtles) and the Brazil Tire cases (rejecting Brazil's import restrictions on certain used tires to reduce negative environmental impacts from tire storage and disposal).

The WTO has a significant degree of lock-in that impedes transition toward a mutually enhancing human-Earth relationship. The approach to the human-Earth relationship reflected in the WTO is largely mimicked in regional trade and investment agreements. The NAFTA approach embodied in the NAFTA and its environmental side agreement and in effect since 1994 is a model with a detailed track record. The USMCA typifies the United States' more recent approach to international trade, including environmental chapters with more details but few if any new binding environmental obligations.

The trade and environment model in NAFTA and subsequent US trade deals

The approach of the United States to international trade and the environment reflects the global approach—with the possible exception of internal trade in Europe, where European Union member countries have accepted supranational authority not only with respect to trade and some aspects of finance, but also with respect to the environment and other domains. The key features of the global trade and finance system are (Rees 2002; Pelletier 2010):

- Its mostly unexamined and increasingly doubtful assumption that mobile capital and limited market regulation lead to long-term social and environmental well-being;
- Its consistent undervaluation of environmental values, biodiversity and ecological integrity;
- Its hard to overcome assumption that measures to protect the environment, public health or other public goods or commons are disguised barriers to trade;
- Its near blindness to overconsumption and aggregate ecological impacts of economic activity; and
- Its assumption that comparative advantage remains an important factor in a world with few meaningful restrictions on the movement of capital.

In light of the momentum this set of features encourages along a path to ecological decline and potential catastrophe, the assumption that they will ensure that boundaries between ecological security and catastrophe will be respected must be questioned. Each element of the institutional infrastructure that supports the current regime of international trade and finance, such as those included in the NAFTA and the NAAEC, should face rigorous scrutiny as to their ability to respect those boundaries.

The United States' approach to trade and the environment is oriented insufficiently toward, and more likely away from, a mutually enhancing human-Earth relationship. The approach in NAFTA and the NAAEC to integrating liberalization of trade with environmental protections, which the United States has continued to pursue with minor changes in the USMCA and other subsequent agreements, increases an ecologically harmful positive feedback. In this feedback loop, trade increases ecological impacts from declining material and energy stocks and rising material and energy flows without countermeasures sufficient to put the global human enterprise on a path toward safe operating space as framed by planetary boundaries and other relevant indicators (Meadows 1999; Rockström et al. 2009). Moreover, international trade and investment rules favor aggregate increases in wealth and profit that are reflected in aggregate measures of economic growth. This approach benefits investors who are far removed from the effects of their economic activity on people and the environment at local scales (Rees 2002; Young et al. 2006; Garver 2019).

At the time of its adoption, promoters of the NAFTA called it the most environmental trade agreement to date (Johnson and Beaulieu 1996; Block 2003). To help get the NAFTA through Congress in 1993, and to meet promises to bolster the modest environmental and labor provisions in the NAFTA text, President Clinton negotiated side agreements with Canada and Mexico on labor and on the environment. The NAAEC was intended to address concerns that liberalized trade in North America would increase environmental impacts due to the increased scale of economic activity; create havens for polluting industries, for example, along the United States-Mexico border; and trigger a "race to the bottom" in which the governments would weaken environmental regulations or enforcement in order to attract economic benefits of trade (Johnson and Beaulieu 1996; McKinney 2000). The NAAEC came into effect on 1 January 1994. It created the trinational Commission for Environmental Cooperation (CEC), headed by a Council comprised of the top environmental official of each country, and imposed on the NAFTA governments obligations designed to minimize environmental concerns. Together the NAFTA and the NAAEC contain provisions to address these concerns.

In its first 20 years, the CEC produced a sustained cooperative program that has yielded some important results (Vaughan 2004). The greatest success came in early years when, for example, the CEC assisted Mexico in phasing out DDT (Commission for Environmental Cooperation 2003a) and began issuing its annual *Taking Stock* report, which combines pollutant, release and transfer information from the three countries. As of April 2020, the citizen-driven submissions on enforcement matters (SEM) mechanism had received 98 submissions and generated 23 factual records (described further below). Sixteen of the factual records were published prior to 2004, and generally, the impact of factual records has been quite limited (Johnson et al. 2004; Garver 2008; Garver 2015).

Overall, measured against its potential to foster aggressive policies on pressing regional and global environmental challenges, particularly related to energy and climate change, the CEC has fallen short (Vaughan 2004; Garver 2008; Garver 2015). Several important environmental provisions of the NAFTA and the NAAEC have been neglected, abandoned or weakened, casting doubt on the environmental promise and legacy of the agreements. Yet some of these weak or neglected provisions have become boilerplate elements in the environmental frameworks negotiated in post-NAFTA trade deals to which the United States has signed on. These post-NAFTA environmental measures raise the concern that they provide "green" cover for trade deals that do little in practice to confront the global ecological challenges that have been increasing—while also becoming ever more apparent—as regional and global trade has expanded (Pollard et al. 2010).

The failure to implement three prominent environmental mechanisms of the NAFTA and the NAAEC and the weakness of a fourth provide evidence of the inability of the NAFTA environmental regime to steer trade toward a mutually enhancing human-Earth relationship. As for the first instrument, the three governments have completely ignored the mild mandate to prevent weakening of environmental protections in the NAFTA countries, to which the States agreed most explicitly in NAFTA Article 1114(2). In regard to the second, Canada, Mexico and the United States have failed to conclude the agreement on transboundary environmental impact assessment (TEIA) that is called for in the NAAEC Article 10(7). The third, the Party-to-Party dispute resolution process set out in Part V of the NAAEC, which aims to provide a remedy for persistent patterns of weak enforcement of environmental laws, has never come to life. The fourth mechanism, related to but independent of the process in Part V, is the modest and largely ineffective submissions on enforcement matters process under Articles 14 and 15 of the NAAEC.

Environmental provisions of NAFTA

The preamble to the NAFTA states the resolve of the NAFTA governments to "undertake [their NAFTA obligations] in a manner consistent with environmental protection and conservation; ... [to] strengthen the development and enforcement of environmental laws and regulations; and [to] promote sustainable development." The text expands on these objectives.

In the event of an inconsistency with the NAFTA, Article 104 of the NAFTA gives qualified precedence to several international environmental agreements. The NAFTA also affirms the Parties' right to adopt environmental standards according to their desired levels of environmental protection, and to impose those standards on other Parties' goods and services, so long as doing so advances a "legitimate objective" and is non-discriminatory (NAFTA Article 904). Further, the NAFTA requires the Parties to "work jointly to enhance the level of safety and of protection of human, animal and plant life and health, the environment and consumers" and to work toward making their environmental standards compatible without weakening them (NAFTA Article 906). A dispute resolution process in Article 20 of the NAFTA allows Party-to-Party disputes regarding the standard-setting provisions and consistency with multilateral environmental agreements.

The most explicit provision in either the NAFTA or the NAAEC that prohibits backsliding, or regression, in levels of environmental protection—albeit softly—is the NAFTA's Article 1114(2). It states:

> The Parties recognize that it is inappropriate to encourage investment by relaxing domestic health, safety or environmental measures. Accordingly, a Party should not waive or otherwise derogate from, or offer to waive or otherwise derogate from, such measures as an encouragement for the establishment, acquisition, expansion or retention in its territory of an investment of an investor. If a Party considers that another Party has offered such an encouragement, it may request consultations with the other Party and the two Parties shall consult with a view to avoiding any such encouragement.

This, along with other provisions in the NAFTA and the NAAEC that reinforce the non-regression obligation, is the first of the neglected provisions; it is examined in more detail below.

Key provisions of the NAAEC

The NAAEC institutionalized a cooperative program for environmental protection and conservation in North America in the CEC and created new mechanisms to enhance public participation in the achievement of North American environmental objectives. Two key measures included in the NAAEC to promote public participation in the work of the CEC are the provision for a process to receive citizen submissions on enforcement in Articles 14 and 15, and the creation of the 15-member Joint Public Advisory Committee (JPAC), which engages the North American public in order to fulfill its mandate to advise the CEC Council on the implementation of the NAAEC.

In regard to enhancing levels of environmental protection without regression, the preamble of the NAAEC reaffirms "the importance of the environmental goals and objectives of the NAFTA, including *enhanced* levels of environmental protection" (emphasis added). This principle reappears among the objectives of the agreement set out in Article 1, the first of which is to "foster the protection and improvement of the environment in the territories of the Parties for the well-being of present and future generations." This implied principle of non-regression is carried forward in the obligations of the NAAEC Parties in Article 3, which states:

> Recognizing the right of each Party to establish its own levels of domestic environmental protection and environmental development policies and priorities, and to adopt or modify accordingly its environmental laws and

> regulations, each Party shall ensure that its laws and regulations provide for high levels of environmental protection and shall strive to continue to improve those laws and regulations.

Clearly, however, the obligation to "strive to continue to improve" environmental laws and regulations is only an implied obligation not to weaken them, and even so, the obligation would be stronger if it was simply "to improve" them.

Other provisions of the NAAEC provide additional reinforcement of the weak non-regression obligation in the NAFTA and the NAAEC. Article 10(3)(*b*) states that

> [t]he Council shall strengthen cooperation on the development and continuing improvement of environmental laws and regulations, including by … without reducing levels of environmental protection, establishing a process for developing recommendations on greater compatibility of environmental technical regulations, standards and conformity assessment procedures in a manner consistent with the NAFTA.

Article 10(6)(b) gives the CEC Council a direct role in the implementation of NAFTA Article 1114(2), requiring it to

> provid[e] assistance in consultations under Article 1114 of the NAFTA where a Party considers that another Party is waiving or derogating from, or offering to waive or otherwise derogate from, an environmental measure as an encouragement to establish, acquire, expand or retain an investment of an investor, with a view to avoiding such encouragement.

To allay concerns that Canada, Mexico or the United States would use weak environmental enforcement to gain economic advantages, Article 5 of the NAAEC requires "each Party [to] effectively enforce its environmental laws and regulations" "with the aim of achieving high levels of environmental protection and compliance with its environmental laws and regulations." Article 5 implies that weakening enforcement would be a form of environmental regression. The SEM mechanism in Articles 14 and 15 and the Party-to-Party dispute resolution process in Part V subject the Party's enforcement obligation in Article 5 to independent review. Part V and the SEM mechanism are discussed in detail below.

The NAAEC also put into effect provisions regarding environmental impact assessment (EIA), an important tool for considering holistically the environmental implications of economic development, including cumulative effects and other impacts that can contribute to providing an overview of the scale effects of economic activities (that is, more trade-stimulated economic activity means more impact). Article 2(1)(*e*) obligates each Party to "assess, as appropriate, environmental impacts." Article 10(6)(*g*) requires the CEC Council to "consider[on an ongoing basis the environmental effects of the NAFTA." Article 10(7) states:

Recognizing the significant bilateral nature of many transboundary environmental issues, the Council shall, with a view to agreement between the Parties pursuant to this Article within three years on obligations, consider and develop recommendations with respect to:

a. assessing the environmental impact of proposed projects subject to decisions by a competent government authority and likely to cause significant adverse transboundary effects, including a full evaluation of comments provided by other Parties and persons of other Parties;
b. notification, provision of relevant information and consultation between Parties with respect to such projects; and
c. mitigation of the potential adverse effects of such projects.

NAFTA Article 1114(2): an impotent shield against environmental backsliding

NAFTA's Article 1114(2) and other provisions noted above strongly imply that environmental backsliding runs counter to the spirit and letter of both the NAFTA and the NAAEC. Differences in levels of environmental protection in the three countries have not caused a clear "race to the bottom" (Block 2003; Wold 2010). Examples of environmental backsliding do exist, however, and illustrate the weakness of the non-regression provisions in the NAFTA and the NAAEC (Moreno et al. 1999). The failure of the NAFTA governments to formally raise concerns regarding such cases has left the potential of NAFTA Article 1114(2) to reinforce environmental protections—as limited as they are—untapped and undeveloped.

Examples of post-NAFTA environmental backsliding

In Canada, the most prominent examples of the weakening of environmental protection were adopted as part of the federal budget implementation acts of 2009, 2010 and 2012, in which significant rollbacks of environmental law were buried deep within multi-faceted annual budgetary and fiscal measures, which were enacted with little public scrutiny or debate. The most severe of these, the *Jobs, Growth and Long-term Prosperity Act*, and the *Jobs and Growth Act*, 2012, which together implemented the 2012 federal budget, plainly weaken Canadian environmental protections in several ways. Most notably, these Acts: (1) amended the federal *Species at Risk Act* (SARA) by eliminating the five-year term limit on permits to engage in activities that affect species protected by SARA or their critical habitat, and exempted certain pipeline projects from the requirement to respect reasonable measures to protect critical habitat of species protected under SARA, (2) replaced the *Canadian Environmental Assessment Act*, 1992, with the *Canadian Environmental Assessment Act*, 2012, which greatly reduced the number of projects subject to federal environmental assessment, reduced the number of factors to be considered in conducting assessments, shortened the time for completing environmental screenings and full environmental assessments and limited opportunities for public participation by tightening the criteria for members of the public wishing to participate in a federal environmental assessment and (3) weakened the federal *Fisheries Act* by increasing the level of harm to fish or fish habitat that would need to be shown in order to prove a violation of the Act, reducing the kinds of fisheries subject to the Act and expanding the ability of the federal government to exempt activities or undertakings from the Act even if they cause harm to fish or fish habitat.

The *Jobs and Economic Growth Act*, 2010, which implemented the 2010 budget, weakened several provisions of the *Canadian Environmental Assessment Act*, 1992, for example, by giving the federal government increased discretion to limit or eliminate the application of EIA requirements for certain projects and activities (Lindgren 2010). The *Budget Implementation Act*, 2009, weakened the scope of the *Navigable Waters Protection Act* by limiting the types of activities and projects that require EIA and other review.

Canada's abandonment in 2011 of its obligations to reduce emissions of greenhouse gases under the Kyoto Protocol on climate change and its related repeal in 2012 of the *Kyoto Protocol Implementation Act* are further examples of backward movement on environmental protection—especially when Canada later adopted greenhouse gas emissions targets for 2020 that were less ambitious than Canada's Kyoto obligations (Government of Canada 2010). Likewise, Canada's reduction of its federal environmental spending by roughly 40 percent in real dollars in the period 1993 to 1997 could itself be seen as a form of regression (Vaughan 2004). Since a change in the federal government in 2015, some but not all of Canada's environmental laws that were weakened have been reinforced again.

In Mexico, one example of regression in environmental protection since 1994 was the dilution in 2004 of a regulation on the protection of wetlands and mangroves (Flores-Verdugo et al. 2007). Noting that environmental improvements in Mexico that some expected NAFTA to produce have not occurred, one set of commentators contended

> Mexico's poor environmental record is mostly due to a weakening of the commitment to environmental protection in the post-NAFTA period. Indeed, real spending and inspection levels in manufacturing have declined since NAFTA, and the environmental impacts have been well documented.
>
> (Zepeda et al. 2009)

Prominent backsliding examples from the United States date back to at least 1995, with the passage of the *Emergency Supplemental Appropriations for Additional Disaster Assistance, for Antiterrorism Initiatives, for Assistance in the Recovery from the Tragedy that Occurred at Oklahoma City, and Rescissions Act*, 1995. That law created a salvage timber harvest program that reduced or eliminated the application of the *National Environmental Policy Act* and other environmental laws to the logging of certain salvage timber sales and some specific controversial timber sales in the Northwestern United States (Moreno et al. 1999). More recent examples include the repeal in 2003 of a rule that had defined total maximum daily loads of water pollution under the *Clean Water Act*, the weakening in 2003 of the regulations on the types of modifications of major air pollution sources that would require more stringent pollution controls under the Clean Air Act's "New Source Review" provisions and the rollback in 2008 of the Stream Buffer Zone rule that prohibited surface mining activities within 100 feet of streams (73 Federal Register 75814 (2008)). A rash of environmental rollbacks has taken place in the Trump Administration, including withdrawal of the United States from the Paris Agreement on climate change, repeal or weakening of the Obama Administration's vehicle emissions standards and pollution standards for power plants and more (Samet and Burke 2020).

One example of post-NAFTA weakening of environmental law in the United States underscores one of the key false promises of NAFTA: that it would open a new era of North American integration in which new opportunities in Mexico would reduce the incentive to migrate north. The 1996 *Illegal Immigration Reform and Immigrant Responsibility Act* authorizes the Secretary of the Department of Homeland Security to forego all legal requirements that apply to the construction of physical barriers (for example, a fence or wall) and roads in the border zone between the United States and Mexico. Between 2005 and 2008, the Secretary exercised this power five times (70 Fed. Reg. 55622 (2005), 72 Fed. Reg. 2535 (2007), 72 Fed. Reg. 60870 (2007), 73 Fed. Reg. 19077 (2008), 73 Fed. Reg. 19078 (2008)) in order to bypass important environmental requirements that otherwise applied to the construction of fences in California and Texas along a significant portion of the United States-Mexico border. These waivers began again in the Trump Administration, in connect with its construction of a wall at the United States-Mexico border.

The failure to implement NAFTA Article 1114(2) and related non-regression provisions

The Parties' conduct since the NAFTA and the NAAEC took effect confirms the weakness of the agreements' non-regression provisions. Commentators have characterized NAFTA Article 1114(2) as a statement of "common policy" that is not mandatory (Gaines 2003; Vaughan 2004). The non-regression provision in Article 3 and most of the other Party obligations in the NAAEC have been described as being "akin to unilateral declarations of intention since they are not enforceable under the NAAEC" (Johnson and Beaulieu 1996: 146). Neither the NAFTA nor the NAAEC contains a reliable means of enforcing the non-regression provisions; non-binding Party-to-Party consultations are the sole remedy under NAFTA Article 1114(2). Thus, the NAFTA governments have weakened environmental protections with no consequences pursuant to the NAFTA or the NAAEC. No Party has initiated consultations regarding regression of environmental protections, and the rollback of environmental laws and regulations has never been on the agenda of the CEC Council, despite the mandate they are conferred in NAAEC Article 10(6)(b). No evidence is apparent that any of the Parties has a program for monitoring compliance with the NAFTA Article 1114(2). Further, the Parties have established neither criteria indicating what qualifies as regression of environmental protections or procedures, nor guidelines for consultations or other means for enforcing the non-regression provisions.

In short, enforcement of those provisions has never been part of the CEC's agenda, and it is difficult to imagine the Parties departing from their record of inaction. The Parties' triple roles as members of the CEC Council, which oversees the NAAEC; as governments subject to the NAAEC's provisions; and as partners in the CEC's cooperative programs create inherent conflicts and disincentives in the enforcement of NAFTA Article 1114(2) and related NAAEC obligations that make progress on meaningful implementation of the non-regression provisions a formidable challenge (Garver 2008; Markell 2010).

Whether the examples of environmental backsliding mentioned in the previous Section are Treaty violations undoubtedly raises issues of interpretation under NAFTA Article 1114(2) or the various provisions of the NAAEC that reinforce the non-regression obligation. In Article 1114(2), what kinds of environmental measures are considered relevant, what is the meaning of the terms "waive" and "otherwise derogate from," and what kind of "encouragement for the establishment, acquisition, expansion or retention in its territory of an investment of an investor" must be shown? Given the very soft nature of the sole remedy of consultation, one would expect these terms to be given a broad and liberal meaning. Certainly, the Canadian prime minister's statements to Latin American investors in April 2012 that touted recent changes to Canadian environmental law as a reason to invest in resource development in Canada would seem to qualify squarely as the kind of encouragement that Article 1114(2) was intended to dissuade (Vizcaino 2012). However, the government Parties—both individually and together as the CEC Council—have done nothing to provide clarity on these questions.

NAAEC Article 10(7): a lost opportunity to expand holistic environmental impact assessment

Article 10(7) of the NAAEC obligates the CEC Council to develop recommendations on various aspects of TEIA within three years of the effective date of the NAAEC's entry into force with a view to an agreement. Article 10(7) derives from the notion in Principle 2 of the 1992 Rio Declaration on Environment and Development—building on Principle 21 of the 1972 Stockholm Declaration on the Human Environment—that "States have … the responsibility to ensure that activities within their jurisdiction or control do not cause damage to the environment of other States or of areas beyond the limits of national jurisdiction." The preamble to the NAAEC explicitly reaffirms this principle. More than 20 years on, a North American TEIA agreement has not been finalized (Garver and Podhora 2008).

Negotiation of a TEIA draft text began after the CEC Council instructed negotiators in 1995 to be guided by overarching principles for TEIA that emphasized the transboundary nature of pollution, good neighborliness, respect for existing national and subnational processes, cost effectiveness and efficiency, complementarity and public participation (Commission for Environmental Cooperation 1995). These overarching principles formed the basis of the Parties' efforts from 1996 to 1999 to negotiate the recommended draft text of a TEIA agreement. The CEC Secretariat facilitated the negotiations by providing background analysis and by hosting, at its Montreal headquarters, meetings of a trinational working group of Party delegates. Consistent with the text of Article 10(7), the negotiations focused on Party-to-Party notification of projects and provision of relevant information, the actual assessment of transboundary environmental impacts, consultations between Parties regarding any conflicts in regard to transboundary impacts and mitigation of adverse impacts.

The challenge under the NAAEC has been to create reciprocal TEIA obligations using existing domestic requirements on EIAs. In June 1999, despite the completion of a nearly complete draft text (Commission for Environmental Cooperation 1997), the negotiations broke down because of difficulty establishing reciprocity on the treatment of projects subject to approval only by subnational (as opposed to federal) governments. In a June 1999 communiqué, the CEC Council seemed to indicate a shift to " 'good neighbor' agreements based on reciprocity" between subnational governments, such as individual states or provinces (Commission for Environmental Cooperation 1999).

Since 1999, progress toward resuming negotiation of a TEIA agreement slowed and then stopped altogether, and the role of the CEC diminished. In its communiqué following the annual Council session in 2001, the Council stated simply that "[w]ith respect to transboundary environmental impact assessment, an informal, productive process is continuing" (Commission for Environmental Cooperation 2001). In a March 2003 letter to the JPAC that makes the lack of movement clear, the Council stated:

> Although substantial progress has been made in the development of the TEIA agreement, the Parties have had difficulty in agreeing on the manner in which the Agreement would address projects subject to approval at the sub-Federal level, due to differences in the Federal systems of each country. Government to government contacts are continuing to explore potential solutions to this problem.
>
> (Commission for Environmental Cooperation 2003b)

In 2004, funding for support of implementation of Article 10(7) was eliminated entirely from the CEC budget.

The three governments subsequently announced that under the Security and Prosperity Partnership (SPP), which Canada, Mexico and the United States undertook in 2005, they would "[s]eek to conclude a transboundary environmental impact assessment cooperation agreement for proposed projects by June

2007" (Security and Prosperity Partnership 2005). The SPP set out a policy framework for North American security and prosperity but had no formal connection to the NAFTA, the NAAEC or the CEC. TEIA negotiations made little progress through the SPP. Neither the Council nor the secretariat has proceeded with any activities related to TEIA since that time.

Part V of the NAAEC: a Pandora's box that will never be opened

Part V of the NAAEC (Articles 22 to 36) targets widespread or systemic failures to effectively enforce environmental laws in the NAFTA countries and establishes Party-to-Party dispute resolution with potential sanctions as the remedy. Article 22 provides that "[a]ny Party may request in writing consultations with any other Party regarding whether there has been a persistent pattern of failure by that other Party to effectively enforce its environmental law." If consultations fail to resolve the matter, the Council can appoint an arbitral panel to prepare a report and, if necessary, assess a "monetary enforcement assessment." Article 36 provides that failure to pay a monetary enforcement assessment can lead to suspension of benefits under the NAFTA. When the NAAEC was negotiated, Part V was seen as the teeth sought by environmentalists to hold the governments to account in the event of weak environmental enforcement (Vaughan 2004).

None of the NAAEC Parties has ever initiated a Part V dispute, and none of them has a discernible internal process for monitoring compliance or pursuing investigations into possible non-compliance by the other Parties. Over 20 years after the NAAEC was negotiated, the Parties have not adopted rules of procedure for Part V, as required under Article 28, and they have not developed a roster of arbitrators, as called for under Article 25. As time progresses, Part V looks more and more like a set of "false teeth" included in the NAAEC in order to ensure the NAFTA's approval by the United States Congress (Vaughan 2004; Garver 2008). There are good reasons to believe that the NAAEC governments will never put Part V to use because of political dynamics built into the structure of Part V and because of past recommendations that a moratorium be placed on its implementation.

The structural obstacles to implementation and use of Part V are significant. NAAEC defines a "persistent pattern" for the purposes of Article 22 as "a sustained or recurring course of action or inaction beginning after the date of entry into force of this Agreement." The agreement also provides that:

> A Party has not failed to "effectively enforce its environmental law" … where the action or inaction in question by agencies or officials of that Party:
>
> (a) reflects a reasonable exercise of their discretion in respect of investigatory, prosecutorial, regulatory or compliance matters; or
> (b) results from *bona fide* decisions to allocate resources to enforcement in respect of other environmental matters determined to have higher priorities.

To succeed in a Part V dispute, a Party would have to overcome these vague definitions in establishing its case and fending off potential defenses and would also have to show that the alleged failure to effectively enforce environmental law

> relates to a situation involving workplaces, firms, companies or sectors that produce goods or provide services:
>
> (a) traded between the territories of the Parties; or
> (b) that compete, in the territory of the Party complained against, with goods or services produced or provided by persons of another Party.

Consequently,

> the outcome will be much less certain than in a typical trade dispute, with completely different stakes and a greater potential for backfiring. [Further,] every plaintiff in these dispute resolution processes is also a potential defendant in a future case, and therefore has a reduced incentive to develop the strongest possible arguments favoring the plaintiff position. It is hard to imagine a government arguing for an expansive definition of what qualifies as a failure to effectively enforce, or for a limited interpretation of defenses that apply to the reasonable exercise of enforcement discretion and bona fide allocation of resources to higher priorities.
>
> (Garver 2008: 39)

For all these reasons, dispute resolution under Part V of the NAAEC "is a Pandora's box no government is likely to open" (Garver 2008: 39).

The SEM process: a modest accountability mechanism with declining impact

Another reason that Part V is not likely to be implemented is fears that the potential for Part V dispute proceedings renders the SEM mechanism under the NAAEC's Articles 14 and 15 less effective (Johnson et al. 2004; Vaughan 2004). Article 14 allows any person residing or nongovernmental organization established in North America to file a submission with the secretariat of the CEC asserting that one of the NAAEC Parties is failing to effectively enforce its environmental laws. The secretariat can ask the targeted Party to respond to a submission and then recommend the development of a factual record to the Council. With a majority vote of the Council, the secretariat conducts a detailed investigation into the allegations and produces a factual record, which requires a majority vote of the Council for publication. A factual record provides information that allows members of the public to decide if the Party has failed to effectively enforce its laws, but the factual record may not provide a conclusion of its own, make recommendations or require a remedy.

The SEM process has a long history of controversy, linked mostly to the NAAEC Parties' dual roles as members of the Council voting on the preparation and publication of factual records and overseeing the process, and as individual Parties targeted by the process (Garver 2008). Perhaps as a consequence, contrary to the apparent purpose of the process, the Council has restricted the ability of submitters to raise widespread, systemic cases of weak enforcement that are the types of assertions most likely to reveal a weak regime of environmental protection (Knox and Markell 2012). In voting to authorize factual records, the Council has on several occasions either limited the scope of the factual records to isolated incidents that were not the main focus of the submissions, as with the Migratory Birds case (involving allegations that the US has never enforced its Migratory Birds Treaty Act against logging operations), or required submitters to provide specific evidence for all cases involved in a widespread or systemic pattern of weak enforcement (Block 2003; Garver 2008).

As a result of the limitations that the Council has imposed on the scope of factual records and other aspects of the process (Garver 2008), the SEM mechanism has not lived up to its potential. Moreover, the governments' mostly tepid responses to factual records to date (Johnson et al. 2004; Garver 2008) suggest the process is not a particularly promising means to hold the Parties to account for weak enforcement. The NAAEC does not require a government that is the subject of a factual record to take any action or to respond in any way following its publication. After stating that follow-up to factual records is a matter of domestic policy of the individual governments (Commission for Environmental Cooperation 2002), the Council more recently adopted measures that merely "call for Parties to follow up on concluded submissions with information on any new developments and actions taken regarding matters raised in such submissions" (Commission for Environmental Cooperation 2012). Overall, the impact of the submissions process and factual records has been modest, with some limited positive effects in specific cases (Johnson et al. 2004; Garver 2008). Factual records in the Migratory Birds case and Pulp and Paper case (alleged weak enforcement of Canada's Fisheries Act and Pulp and Paper Effluent Regulations against pulp and paper mills in Eastern Canada) had no discernible impact on the enforcement practices of the government concerned.

Conclusions regarding NAFTA and the NAAEC

Global and regional ecological challenges related to climate change, biodiversity loss, excessive human introduction of nitrogen and phosphorous, overfishing and the use of land and freshwater have steadily grown more pressing since the NAFTA and the NAAEC took effect, along with awareness of those impacts (Speth 2008; Rockström et al. 2009; Carpenter and Bennett 2011; Steffen et al. 2015). International trade is increasingly a driver of these challenges (Hails et al. 2008). These trends make clear the urgency of developing strong measures to prevent backsliding in environmental protection, to ensure effective enforcement and—most critically—to counter the ecologically negative scale effects of expanded economic activity. The ongoing risk of economic crises like the global recession of 2008 and 2009, which can tempt governments to roll back environmental regulation or enforcement and to expediently promote environmentally damaging development in order to stimulate economic growth and job creation, underscores the need for such protections.

The NAFTA and the NAAEC are not suitable occupants of their policy space, and they do not give governments the tools or an enforceable obligation to drive a transition to an economy that lives within ecological limits. Environmental mechanisms in NAFTA and the NAAEC were adopted to provide key guarantees that liberalized trade would promote stronger environmental protection and enforcement and improved analysis of environmental impacts to address scale effects. Not only have these been ignored or suppressed in the context of the NAFTA and the NAAEC, but versions of the NAFTA's Article 1114(2) have also come to appear in virtually every subsequent trade deal that the United States and Canada have entered into,[1] and versions of Part V of the NAAEC appear in the USMCA and most other subsequent United States trade agreements.[2] These empty provisions take up policy space that might otherwise be occupied by effective innovations able to address the aggregate ecological effects of an increasingly globalized economy, grounded in a focus on moving toward and preserving a mutually enhancing human-Earth relationship.

The NAFTA governments know how to adopt more rigorous tools if they want to. They did so in the NAFTA's Chapter 11 by waiving their sovereign immunity to allow private investors to pursue binding arbitration seeking judicially enforceable monetary awards that include estimated lost profits for breach of the NAFTA's investor protections. Although a comprehensive review of Chapter 11 cases involving challenges to environmental measures is beyond the scope of this analysis, two recent cases bear mention.

In March 2015, a Chapter 11 panel of three arbitrators in *Clayton/Bilcon v. Government of Canada* ruled two against one that Canada violated Chapter 11 as a result of a joint federal-provincial environmental assessment process that led to Canada's rejection of a marine terminal on the Digby peninsula in Nova Scotia that a United States-based company sought to ship basalt from a quarry in the area to the United States. The federal and provincial governments concluded that the project "poses the threat of unacceptable and significant adverse effects to the existing and future environmental, social and cultural conditions influencing the lives of individuals and families in the adjacent communities" (Parent 2007). In January 2019, the arbitral panel awarded the investors US$7 million plus interest.

In *Lone Pine Resources Inc. v. Government of Canada*, a United States investor is claiming nearly US$120 million to compensate for the Quebec government's revocation under its 2011 *Act to limit oil and gas activities* of the investor's licenses to explore for oil and natural gas along and near the St. Lawrence river near Trois Rivières, Quebec. The investor anticipated that the exploration licenses could lead to shale gas development in the region. Based on a strategic assessment of the impacts of shale gas and other hydrocarbon development on the human and biophysical environment, the government of Quebec concluded that the region was not suitable for hydrocarbon development.

Table 14.1 summarizes the key differences between the SEM process and the investor dispute mechanism in NAFTA Chapter 11. Similar IDS mechanisms have been included in all United States trade agreements since NAFTA, including

Table 14.1 Comparison of SEM process in NAAEC Articles 14 and 15 and Investor-State Dispute mechanisms in NAFTA Chapter 11

	SEM Process	*NAFTA Chapter 11*
Who can file?	Non-governmental organizations of persons residing in a NAFTA country, asserting failure to effectively enforce environmental law.	Investors of another Party, seeking compensation for expropriation of investment or other losses due to violation of NAFTA rules.
Description of process	Submission filed with Secretariat; Council must approve detailed investigation and release of results.	Investor submits claim to arbitration; NAFTA include Parties' consent to arbitration; arbitration pursuant to ICSID (International Centre for Settlement of Investment Disputes) or UNCITRAL (United Nations Commission on International Trade Law) rules.
Role of complainant	Files submission; may provide information during factual record development.	Selects one arbitrator and may object to third arbitrator; full party to arbitration; may submit evidence and argument and respond to government.
Authority of tribunal	Secretariat recommends factual record, for Council vote; no subpoena power or authority to take testimony under oath.	Arbitral panel has authority to hold evidentiary hearings, issue orders, allow *amicus curiae* briefs and make binding awards.
Transparency	The CEC posts all determinations in the process, as well as factual records, on its website. Parties can keep confidential some kinds of information and can vote not to publish factual records.	Applicable arbitration rules were designed primarily for commercial arbitration with limited transparency. Public participation is generally excluded. NAFTA governments have agreed to make proceedings publicly available, with exceptions.
Remedy	Factual record, with no conclusion on whether country failed to effectively enforce and no formal follow-up.	Binding arbitration award compensating investment loss, including lost profits, typically multi-million-dollar awards where successful.

the pending Transatlantic Trade and Investment Partnership (TTIP), except the USMCA.

The dispute settlement mechanism in Part V of the NAAEC, submissions on enforcement matters under the NAAEC's Articles 14 and 15 and the environmental consultations otherwise allowed under the NAAEC or the NAFTA are weak cousins to the strictly enforceable investor dispute mechanisms in the NAFTA and subsequent trade deals. In light of the contribution of regional and global trade to looming ecological challenges like climate change, this differentiation in the power to access publicly financed remedies reflects a troublesome imbalance. Without a meaningful attempt to hold governments to account in regard to environmental backsliding or other environmental shortcomings, or to give critical issues like climate change paramount importance in regional decision-making, the perception that the environmental provisions in trade agreements are mere paper tigers against North American ecological challenges is likely to continue and to solidify.

Overcoming the ecological stagnancy of trade policy in North America faces resistance because of the difficulty in transforming institutions that have achieved stability, even where the conditions and assumptions under which those institutions were established have changed significantly or have been shown to be faulty (Daniels 2007). In this case, the relevant institutions are an entire series of trade agreements, from NAFTA to the USMCA, built around an approach to trade and environment that lacks a holistic, systems-based framework to ensure progress toward a mutually enhancing human-Earth relationship. The examples presented above shed light on the bureaucratic and administrative minutiae that maintain this stagnant situation. This phenomenon of maintaining "tragic institutions" (Daniels 2007)—that is, a lock-in tendency by which institutions created to address environmental concerns become ineffective but difficult to change—is exacerbated where pressures in addition to the bureaucratic tendency to maintain budgets and programs are considerable. In the case of the North American approach to trade and the environment, a key factor is the expediency of reproducing a model when doing so does not appear to come with any significant political cost.

Institutions established to implement and enforce legal mandates are essential, but "with a change of circumstances and values sensible institutions can morph into tragic institutions" (Daniels 2007: 565). Tragic institutions become unable to meet the objectives for which they were created and yet resist replacement or reform (Daniels 2007), and often create "rulemaking ruts" in which rules resist change in response to new information (Blais and Wagner 2008: 1738). In the context of trade and environment in North America, a key change in circumstances since 1994 is the significant post-NAFTA evidence regarding the accelerating regional and global ecological crisis, especially climate change. Yet, the trade-based institutional infrastructure for addressing this crisis can only be described as a tragic institution. According to Daniels (2007: 566–568), institutions are more likely to be responsive and adaptive if they have the capacity to (1) promote public participation and transparency, (2) prepare those who control commons resources to be flexible and adaptive, (3) be "continually expos[ed] to competing values," (4) "[w]ithin sensible bounds, allow trading among users," (5) "[b]uild mechanisms to internalize externalities," (6) "[p]rovide incentives for users to conserve the commons," (7) give legal rights to those affected by the use of commons to challenge decisions regarding management of the commons and (8) "[i]f necessary, buy out interests of entrenched users." With these criteria in mind, a more fundamental reform of the United States' approach to trade and the environment is needed—one that reckons with aggregate scale effects of the regional and global trade and finance regime in radical new ways. The USMCA does not make these radical changes.

Transition to an ecologically bounded legal regime for trade in the Anthropocene

A legal regime that accords with global ecological boundaries and a mutually enhancing human-Earth relationship would impose on human activities limitations that do not exist under the current legal systems in the developed world. A system in which ecological restraints envelop the economy requires the collective of economic actors to limit their choices so that, taken together, they respect those ecological limits. Yet, in market economies, the freedom to spend on activities and maximize personal wealth, often regardless of their ecological costs, is paramount (Rees 2002). Legally protected consumption, rooted in strong notions of property rights and personal freedom, exacts ecological costs that are collectively shared and often incremental, diffuse and delayed and consequently mostly external to the economic and legal infrastructure. Analyses of the trade flows of material and energy resources indicate that their consumption in high-income countries contributes significantly to biodiversity losses and other ecological harms that are most severe in low- and middle-income countries (Srinivasan et al. 2008; Pollard et al. 2010).

On the flip side, the economic model that current international trade and finance regimes seek to extend and globalize accords strong private property rights to the owners of production (Beddoe et al. 2009). Reflecting the core ideals of Locke and Smith, the dominant global economic approach is designed to give the proprietors and sellers of goods, services and intellectual property, including shareholders and investors aiming primarily to maximize their profits, as much freedom as possible to build, innovate, expand and increase their wealth. At all scales, environmental and other restrictions have been imposed to protect public and common interests in too piecemeal and too atomized a manner to adequately balance the global human-Earth system. Moreover, within this infrastructure, stronger intellectual property rights that trade agreements like the NAFTA and the USMCA instill may impede the flow of information that could lead to or expedite solutions to ecological problems (Heller 1998; Beddoe et al. 2009). As such, investors, producers and consumers are all complicit in an expanding global economy that has exceeded some global ecological boundaries and is on course to worsen the ecological dilemma.

The current legal and institutional infrastructure for international trade and finance does little to address these ecological and social stresses. In particular, the WTO system aims to provide a rule-based, predictable and fairly rigid regime for proprietors of goods, services and intellectual property, as well as investors, to attain their economic goals with only piecemeal consideration of relevant ecological factors. A tall order for international trade and investment law in the Anthropocene is to transition toward an adaptive regime in which trade and investment rules are adjusted in response to ongoing monitoring of the global, regional and local impacts on the human-Earth relationship. In fairly distributing responsibilities for staying within global ecological boundaries, these adaptive trade and investment rules must account for trade patterns by which goods and services embody ecological impacts that occur at great distances away.

For example, international trade and the resulting metabolic rift between where materials are extracted or harvested and where they are consumed or disposed of (Clark and Foster 2009), as well as other interdependencies between regions and nations, affect the geographic distribution of climate change drivers and of the fair assignment of responsibility for mitigating or adapting to its effects. Unequal geographic distribution of the benefits and negative impacts of activities that cause climate change, such as sea level rise attributable to climate change that affects places that contribute relatively little to climate change, should also influence governance responses (Srinivasan et al. 2008). The locked-in greenhouse gas emissions related to long-term development patterns, such as fossil-fuel dependent post-World War II infrastructure, technology and product design that would take time, resources and political determination to reconfigure, replace or make obsolete, must also be considered. Human population trends with long-term momentum are another geographic factor with which to contend within this context.

Key metrics and indicators

The US approach to trade and investment fails to integrate a comprehensive approach for monitoring and effectively responding to the aggregate ecological impacts related to the flows of material and energy that international trade and investment facilitate. By contrast, the European Union's strategy for sustainable development takes into account that

> the externalisation of environmental burden through international trade might be an effective strategy for industrialised countries to maintain high environmental quality within their own borders, while externalising the negative environmental consequences of their production and consumption processes to other parts of the world.
>
> (Giljum et al. 2009: 140)

In the Anthropocene, a more rigorous, comprehensive and ecologically grounded system of law and governance than is reflected in US trade agreements is needed to ensure that international trade and investment promotes and preserves a mutually enhancing human-Earth relationship. Linking planetary boundaries and other metrics and indicators with material and energy flows that trade and investment facilitates and with the consequent ecological impacts across spatial and temporal scales is fundamental to such a system of law and governance.

Planetary boundaries

Proposed planetary boundaries (Rockström et al. 2009; Steffen et al. 2015) are the main indicators considered here for accounting for the aggregate global scale of the ecological impacts of international trade and investment. A global legal system is needed with enforceable limits on aggregate ecological impacts, as represented by planetary boundaries and other relevant indicators. Sub-global distribution of allowable contributions toward those limits, for example by the United States, should be determined according to consumption of goods and services that drives pressures on the planetary boundaries. Considerable uncertainties remain as to several of the planetary boundaries as well as to the relationship of material and energy flows, and patterns of consumption and production, to ecological impacts that the boundaries represent. A precautionary approach allows limits to be enforced despite uncertainties—which are inevitable in social-ecological systems at all scales. The rigorous assessment of Sweden's environmental impacts with respect to the planetary boundaries framework, based on an analysis that attributed embedded ecological impacts of imports and exports at the place of consumption (Nyquist et al. 2013), is a good example of how global ecological limits can be apportioned at sub-global scales. It is a hopeful sign that this work is carrying on and becoming more specific, sophisticated and close to being operational over time (Häyhä et al. 2016; Häyhä et al. 2018).

Tracking material and energy flows: the example of HANPP

International trade is "an instrument for transferring large amounts of energy and materials throughout the world" (Gonzalez de Molina and Toledo 2014: 137). For example, in 2000, 20 percent of extracted materials were exchanged in international trade (Gonzalez de Molina and Toledo 2014: 137). Biomass represents a significant portion of this material trade (Krausmann et al. 2009). Net primary production (NPP) is the biomass that plants accumulate through photosynthesis. Metrics of human appropriation of net primary production (HANPP) track the amount of biomass that humans appropriate through harvest, grazing by livestock, shifts in land use or other means (Haberl et al. 2007). In turn, HANPP provides useful information regarding trade-offs between human provisioning from agricultural, settlement and other appropriations of biomass, and ecosystem processes that maintain biodiversity and influence hydrologic and biogeochemical cycles, often in ways that benefit humans (Erb et al. 2009a). For example, research on the impact of HANPP on species extinctions and other aspects of biodiversity related to the planetary boundary for biosphere integrity helps clarify criteria for reducing those impacts (Vitousek et al. 1986; Wright 1990; Haberl et al. 2004; Haberl et al. 2007; Erb et al. 2009a). Haberl et al. (2007) used an extensive set of spatially explicit data to produce a high-resolution (5', or roughly 10 km by 10 km) spatial distribution of HANPP at the point of biomass appropriation. Spatially explicit information on HANPP where it is appropriated is important for assessing the relationship of HANPP to localized biodiversity loss and other heterogeneous impacts of HANPP on ecosystem structure and function (Haberl et al. 2007). Moreover, it provides a more precise picture of human domination of ecosystems than does information on the point of consumption (Imhoff et al. 2004).

Spatially explicit metrics of HANPP allow rigorous analysis of global, cross-scalar patterns and processes, in particular trade, by which biomass moves from the point of appropriation to the point of consumption. Foley et al. (2007) analyzed the fate of HANPP showing the geospatial distribution of the percentage of crop production devoted to food and non-food (feed for livestock, fiber and luxury items such as tobacco) products, and of the percentage of crop production destined for export. Erb et al. (2009b) analyzed embodied HANPP, which accounts not only for the biomass contained in products at the point of consumption, but also the other upstream components of HANPP involved in the production process. They calculated consumption of embodied HANPP by adding domestic HANPP (calculated per Haberl et al. 2007) and imported embodied HANPP and subtracting exported embodied HANPP. Imported and exported embodied HANPP were estimated at the national level by multiplying domestic HANPP values calculated according to Haberl et al. (2007) by country-specific ratios of biomass products to upstream HANPP totals. National totals of embodied HANPP were then disaggregated and mapped using spatially explicit data on population densities, yielding a spatial distribution of consumption of embodied HANPP at the same spatial resolution as in Haberl et al. (2007). Combining these results with the HANPP maps in Haberl et al. (2007) gives a spatially explicit picture of net-consuming and net-producing areas across the globe in 2000. Erb et al. (2009b) show that embodied HANPP tends to flow from countries with low population density to countries with high population density, regardless of development status, but with plenty of underlying complexity.

Calculating spatially explicit values of HANPP, as well as spatially explicit metrics of other kinds of material and energy flows using an input-output methodology, is highly data intensive and requires care to avoid double counting (Haberl et al. 2007; Erb et al. 2009a). Despite these challenges, spatially explicit information on HANPP and the flows of harvested biomass from the points of appropriation to consumption endpoints is important for understanding the drivers and implications for sustainability of agriculture, timber harvest, forest fires, urbanization and other land use change and other forms of biomass appropriation. Tracking the movement of HANPP and other material and energy sources in the global economy at high resolution, in conjunction with geographically explicit data on the drivers and impacts related to climate change, biodiversity, biogeochemical fluxes and other key systemic processes, will provide a strong basis for developing legal systems that promote a mutually enhancing human-Earth relationship. Ongoing extension of the analysis of HANPP can help in particular with challenges related to feeding a rising human population that is trending toward greater consumption of meat; deciding the future role of biofuels in the global energy picture; contending with the trend toward increasing metabolic rift in an increasingly globalized economy and controlling the impacts of intensive biomass production that depends on fertilization, genetically modified organisms, pesticides and irrigation (Erb et al. 2009a; Foley et al. 2011; Bringezu et al. 2012).

Extensions of concepts like embodied HANPP, which tracks biomass from the point of harvest to the point of consumption through trade, "could provide valuable information on sustainability aspects of socio-ecological systems, linking drivers and impacts of land use" (Erb et al. 2009a: 257). The critical loads and levels approach of the Convention on Long-range Transport of Air Pollution and the Montreal Protocol on ozone-depleting substances have both had some success and could serve as models (Garver 2013). Metrics of the amount of non-renewable energy resources, particularly hydrocarbons, and other non-renewable materials, such as metal and uranium, recovered per unit of energy invested, such as EROEI, are also relevant to making well-reasoned rules and policy. EROEI and similar metrics can guide market corrections needed to enable the maximum acceptable extraction of non-renewable materials within ecological limits such as those related to climate change.

This approach counters approaches that accept projections of human consumptive demand (often labeled "needs") based on business-as-usual, market-driven trends uncritically and orient governance toward ensuring that material and energy supplies will meet that demand (Nordhaus 2007; Dobbs et al. 2011). Although analyzing such demand projections is prudent, more rigorous analyses include proposals to reduce ecological impacts by driving down demand, for example by promoting shifts away from meat diets and reducing waste in order to avoid a projected doubling of agricultural production to meet food demand due to population growth, dietary changes and competition from biofuels (Foley et al. 2011). Such shifts may be difficult to achieve socially and politically but labeling innovative options that target demand as unrealistic (Nordhaus and Yang 1996) underestimates the extent of the current ecological predicament—perhaps most clearly with respect to climate change. An approach like UNEP's green economy strategy, which emphasizes green investment in key sectors, may have a transitional role to play, because rapid gains in efficient use of material and energy should be part of any governance scenario. The UNEP Green Economy Report dodges the growth-conundrum bullet by relying on a model that suggests that robust growth is possible in a green economy (at least in the short term), but reasoned analysis has revealed flaws in that model (Victor and Jackson 2012) and other integrated global models point to opposing conclusions (Meadows et al. 2004).

Key leverage points

The current international trade and investment system, typified by the WTO, the NAFTA model and the USMCA, enhances a positive feedback loop in which trade and investment increase aggregate global material and energy flows and associated ecological impacts at the global, regional and local scale. Existing mechanisms to address ecological concerns within this growth-driven approach are a form of system restraint that Meadows (1999) presents as a superficial or at best mid-level means for changing the functioning of a complex system—in this case, the human-Earth system. Yet, because those mechanisms are not based on a holistic, systems-based approach, and are encompassed within a growth-insistent economic model that does not integrate ecological limits, they do not adequately use this leverage point to move toward a mutually enhancing human-Earth relationship. More effective leverage points, based on shifts in paradigm, goals or rules, are missing in the current system.

Leveraging modest environmental restrictions while retaining the higher-level rules within Meadows's hierarchy that drive the current global trade regime is insufficient to advance a mutually enhancing human-Earth. At the level of paradigms, a shift is necessary toward a view of humans as integrated with, and interdependent upon, the ecological foundations of the human enterprise. In regard to goals, a shift is needed from the goal of perpetual economic growth on which the global regime of international trade and finance is founded to a goal of a mutually enhancing human-Earth relationship. In terms of rules, the priority given to trade and investment and enhancement and protection of private property interests must give way to primacy for global ecological limits, distributed regionally and locally, and to greater recognition of the global commons. The emphasis on predictability and rigidity of trade and investment rules must give way to the flexibility and adaptiveness needed to respond to monitoring of the impacts of trade and investment on the human-Earth relationship. Table 14.2 summarizes key differences in how different leverage points can be applied under the current approach as compared to an approach that respects ecological limits.

Lock-in assessment

Many key components of the current global system of international trade and investment, in which the United States is a dominant agent, have a high degree of lock-in that undermines a mutually enhancing human-Earth relationship. Without a system shock or surprise, or unprecedented international cooperation, it will likely take a long time to shift toward an international trade and investment regime with highly locked-in features that promote a mutually enhancing human-Earth relationship. However, some features of the current regime fall within the transitional zone between a growth-insistent paradigm and a limits-insistent paradigm. This transitional zone includes measures that promote trade and investment in goods and services that enhance environmental protection and

Table 14.2 Leverage points: current approach compared to proposed approach for ensuring that trade and investment respects ecological limits

Leverage point (per Meadows (1999))	*Current trade and investment regime*	*Trade and investment regime consistent with a mutually enhancing human-Earth relationship*
12 (parameters)	Material and energy flows related to trade and investment are affected by changes in tariffs and other measures, including the level of protection of foreign investments, burdens of proof in dispute resolution, etc. No system exists for holistically tracking and responding to the ecological impacts of these flows.	Integrated ecological and economic standards related to trade and investment are adjusted within a precautionary and adaptive framework that gives primacy to ecological limits.
11 (buffers)	Stocks of wealth from international trade and investment are potentially available to help pay for a transition toward a mutually enhancing human-Earth relationship, but this occurs only with higher-level changes, such as a significant shift from private to public or common property regimes.	Stocks of material and energy needed to provide a sufficient level of well-being for humans within a mutually enhancing human-Earth relationship distributed spatially; Nations design trade and investment to create shared ecological buffers as needed.
10 (stock and flow structures)	Trade and investment stock and flow structures have well-developed, highly locked-in and interrelated technological, physical, institutional, political and economic structures and features.	Commitment to a mutually enhancing human-Earth relationship leads nations to develop stock and flow structures with entropic thrift (Brown 2015) and fair sharing intragenerationally, intergenerationally and among species.
9 (delays)	Ecological impacts related to enhanced international trade and investment are often delayed; no systems-based approach exists for continually monitoring impacts across spatial and temporal scales and for understanding how trade and investment activity causes them.	A systems-based approach for continually monitoring ecological impacts related to enhanced trade and investment and for understanding how trade and investment activity causes them allows for fine-tuning relevant rules through an adaptive approach.

Leverage point (per Meadows (1999))	*Current trade and investment regime*	*Trade and investment regime consistent with a mutually enhancing human-Earth relationship*
8 (negative feedback)	Some forms of trade and investment enable greater self-sufficiency in some parts of the world, leading to reduced trade; increased trade in environmental goods and services could reduce the intensity of the ecological impacts of economic and financial activity, but not the absolute material and energy flows associated with them.	Trade and investment rules are embedded within a framework that gives primacy to ecological limits; and that includes supranational trade and investment measures that restrict overconsumption, control of local ecosystems by remote owners, inefficient material and energy flows and other things that undermine the human-Earth relationship across scales.
7 (dampen positive feedback)	Liberalized international trade and investment rules enhance a positive feedback that increases material and energy flows and depletes material and energy stocks; piecemeal environmental measures are not comprehensive or well enforced, and therefore do not dampen this positive feedback; enforcement of rules prohibiting environmental backsliding to encourage trade and investment might have a damping effect.	Nations embed measures that promote trade and investment rules within an overarching framework that gives primacy to ecological limits. This yields supranational measures that restrict overconsumption, control of local ecosystems by remote owners, inefficient material and energy flows and other things that undermine the human-Earth relationship across scales.
6 (information flows)	The international community tracks information regarding the human-Earth relationship in a piecemeal, non-holistic manner, with much critical information omitted; information collected has little impact on the overall trade and investment system structure and functioning.	The international community tracks information regarding the human-Earth relationship in a holistic manner and designs the trade and investment system structure to be adaptive in response to this information.
5 (rules)	Rigid, predictable rules promote economic growth, protect private property interests in trade and investment, minimize protections in the public interest and respect state sovereignty except with respect to foreign investments. These rules allow private investors to seek compensation from public funds for violation of supranational investment rules, in arbitration with limited transparency.	The international community adopts a flexible, adaptive system of rules and makes adjustments based on monitoring of impacts of trade and investment on the human-Earth relationship, with primacy to respecting ecological limits.

Continued

Table 14.2 Continued

Leverage point (per Meadows (1999))	*Current trade and investment regime*	*Trade and investment regime consistent with a mutually enhancing human-Earth relationship*
4 (system structure)	The international community integrates the trade and investment system with related political, economic and technological systems at the global scale, with a focus on maintaining growth; there is little flexibility in system structure.	The international community designs an adaptive approach that allows the system structure to evolve in response to research, experience and monitoring.
3 (goals)	Growth insistence; least possible regulation of trade and investment.	Mutually enhancing human-Earth relationship.
2 (mindset)	Humans as dominant over nature; strong private property rights and an open market economy encourage best use of material and energy resources for humans.	Humans as integrated into and dependent upon the ecosphere.
1 (transcend paradigms)	No transcendent approach.	Enhancement of heterogeneous paradigms in socially constructed systems in the global panarchy.

Table 14.3 Summary of key lock-in tendencies that hinder international trade and investment that promotes a mutually enhancing human-Earth relationship

Lock-in challenge	*Degree of lock-in*	*Spatial scale*	*Time for resolution*
Commitment to sustained economic growth	High	Local to global	Long
"Tragic institution" effect: e.g. continued reliance by US & Canada on outdated NAFTA model	Medium-High	Regional to global	Medium-Long
Investments in trade infrastructure: production, transport, ports, information, institutional, financial, etc.	Medium-High	Local to global	Medium-long
Socio-cultural (e.g. consumers prefer less expensive goods and services, even if prices ignore ecological degradation)	High	Local to global	Medium-long
Ecological (e.g. aggregate impacts of trade are often delayed and are dispersed and difficult to monitor in the aggregate)	High	Local to global	Long

Table 14.4 Lock-in assessment: obstacles in US law to international trade and investment that would promote and preserve a mutually enhancing human-Earth relationship

Obstacles (Examples)	*Key leverage points (per Meadows 1999)*	*Nature and importance of social and ecological stakes*	*Degree of lock-in*	*Time to overcome*
Strong US commitment to growth-insistence as a primary goal for trade and investment	1, 2, 3, 4, 5	In a growth-insistent economy, maintaining a growing economy has primacy over ecological limits. Resource productivity that allows perpetual growth is ultimately insufficient to limit greenhouse gas (GHG) emissions and other causes of ecological impacts enough to meet planetary boundaries.	High	Long
Entrenched mechanisms in US laws regarding international trade and finance, mostly derived from the NAFTA model	2, 3, 5, 6	Overall negotiating objectives for trade and investment have remained relatively unchanged since NAFTA, and political pressure from civil society on trade issues has receded (e.g. although some environmental groups continue to oppose the US approach to trade and investment, influential environmental groups have not collectively developed and advocated alternatives to the NAFTA model on trade and environment).	Medium-High	Medium-Long

continued

Table 14.4 Continued

Obstacles (Examples)	*Key leverage points (per Meadows 1999)*	*Nature and importance of social and ecological stakes*	*Degree of lock-in*	*Time to overcome*
Interpretation of the provisions on private property rights in the US Constitution	1, 2, 5	Strong private property rights regimes in US law and policy have strongly influenced protections of the rights of investors in NAFTA and subsequent US trade deals. Investors are likely to strongly resist any changes to existing trade deals, or alternative approaches in future trade deals, that significantly reduce those protections for ecological or other reasons.	High	Medium-Long
Role of the US in the creation and ongoing authority of the WTO	2, 3, 4, 5, 6, 7	The global trade regime is a key component of the global commitment to economic growth, based on conventional economics that inadequately considers systems-based consideration of the ecological impacts of material and energy extraction and flows.	Medium-High	Medium
US resistance to supranational law outside of the trade and finance domain	2, 3, 4, 5	Strong supranational rules in international law exist only at the regional level with respect to the environment (e.g. the EU) and at the global level only with respect to trade and investment. Yet, they are likely essential for staying within planetary boundaries.	Medium-High	Long

Notes
Meadows's leverage points are coded as follows, per Meadows 1999; Figure 4.2: (1) Transcend paradigms, (2) Mindset, (3) Goals, (4) System structure, (5) Rules, (6) Information flows, (7) Dampen positive feedbacks, (8) Negative feedbacks, (9) Delays, (10) Stock and flow structures, (11) Buffers, (12) Parameters.

Table 14.5 Lock-in assessment: opportunities for US law on international trade and investment that fosters and preserves a mutually enhancing human-Earth relationship

Opportunities (examples)	*Key leverage points (per Meadows 1999)*	*Nature and importance of social and ecological stakes*	*Time to implement*
Enforceable global GHG cap, based on a precautionary approach as with the planetary boundaries that would account for trade and investment	2, 3, 4, 5, 7, 10, 12	Adoption and enforcement of a global GHG cap, such as a cap on cumulative emissions and a detailed plan for distributing the right to emit, would mark a significant turning point in addressing climate change.	Medium-Long
Expansion of common/public property rights within US law	1, 2, 3, 4, 5	Expanding common and public property rights in US law, with a corresponding reduction in private property rights, would likely make it easier to impose restrictions or obligations that would mitigate climate change, both in terms of emissions and deforestation. Constitutional protection of private property rights, as well as strong socio-political support for strong private property rights, are a significant obstacle.	Medium-Long
Rational use of planetary boundaries, Energy Return on Energy Invested (EROEI) and related metrics	2, 3, 4, 7	Strong US support for domestic and international commitment to giving planetary boundaries primacy in law and governance could significantly transform global law and governance, including law and governance related to international trade and investment.	Medium-Long

continued

Table 14.5 Continued

Opportunities (examples)	*Key leverage points (per Meadows 1999)*	*Nature and importance of social and ecological stakes*	*Time to implement*
Adequately supported (financially and politically) global and sub-global institutions for continually (1) monitoring impacts of trade on ecological limits, as represented by planetary boundaries, (2) adapting trade and investment rules in response to this monitoring and (3) giving primacy to ecological limits and a mutually enhancing human-Earth relationship in implementing and enforcing trade and investment rules	4, 5, 7, 8, 9	The WTO and US domestic institutions (e.g. the US Trade Representative) created to support international trade and investment demonstrate the ability to create strong, international institutions that implement and enforce a rule-based legal regime linked to supportive national legal and institutional structures. The European Union demonstrates a similar structure with supranational legal authority coordinated with national legal systems—but in the EU, this structure has expanded beyond trade and finance issues to include environmental, labor, health and other domains. The proposals by Brown and Garver (2009) for global law and governance are one possible model. In particular the proposed Global Reserve would comprehensively monitor human impacts on the Earth's life support capacity across spatial and temporal scales using indicators like planetary boundaries, and propose or impose sub-globally distributed limits on human activities so as to promote and preserve a mutually enhancing human-Earth relationship.	Medium-Long

Notes
Meadows's leverage points are coded as follows, per Meadows 1999; Figure 4.2: (1) Transcend paradigms, (2) Mindset, (3) Goals, (4) System structure, (5) Rules, (6) Information flows, (7) Dampen positive feedbacks, (8) Negative feedbacks, (9) Delays, (10) Stock and flow structures, (11) Buffers, (12) Parameters.

conservation of biodiversity, as well as measures that integrate multilateral environmental agreements that include some constraints—if adequately enforced—on trade and investment.

Table 14.3 summarizes key lock-in challenges in the transition toward international trade and finance that promotes and preserves a mutually enhancing human-Earth relationship. Tables 14.4 and 14.5 provide a preliminary and non-comprehensive qualitative assessment of lock-in features of the United States' current legal regime regarding international trade and investment that impede progress toward a mutually enhancing human-Earth relationship, and of the resistance to the adoption of legal mechanisms regarding international trade and investment that would foster and preserve a mutually enhancing human-Earth relationship. The key leverage points according to the hierarchy in Meadows (1999) are indicated for each obstacle or opportunity. This preliminary assessment provides a framework for conducting more detailed and rigorous qualitative and quantitative assessments for the systems features highlighted.

Conclusions

The current global system of laws that govern international trade and finance lacks the holistic, systems-based framework needed for it to contribute to a mutually enhancing human-Earth relationship. The approach presented here of conducting lock-in assessments of the United States' and other countries' approaches to international trade and finance, with a focus on leverage points with the greatest transformative potential, provides such an overarching framework. Analysis of the material and energy flows associated with international trade and finance and how those flows and depletions of key system stocks increase pressures on planetary boundaries across spatial and temporal scales, is integral to this approach. Refining this framework and putting it into effect will require concerted, interdisciplinary policy-focused research, hopefully combined with real-world experiments in a variety of contexts—activist, community-based, policymaking or other. In Part V, I suggest the international degrowth movement as a promising testing ground for the ideas and proposals in this book.

Notes

1 See e.g. Free Trade Agreement between Canada and the Republic of Colombia, Lima, 21 November 2008, in force 15 August 2011, Can. TS 2011 No. 11, Article 815; Free Trade Agreement between Canada and the Republic of Panama, Ottawa, 14 May 2010, in force 1 April 2013, Can. TS 2013 No. 9, Article 9.16; Free Trade Agreement between Canada and the Republic of Peru, Lima, 29 May 2008, in force 1 August 2009, Can. TS 2009 No. 15, Article 809; Free Trade Agreement between the Government of Canada and the Government of the Republic of Chile, Santiago, 5 December 1996, in force 5 July 1997, Can. TS 1997 No. 50, Article G-14; Dominican Republic–Central America–United States Free Trade Agreement, 28 May 2004, (2004) 43 ILM 514, Article 17.2(2); United States–Colombia Free Trade Agreement, 22 November 2006, in force 15 May 2012, Article 18.3(2) [online]: http://ustr.gov/trade-agreements/

free-trade-agreements/colombia-fta/final-text; United States–Peru Trade Promotion Agreement, 12 April 2006, in force 1 February 2009, Article 18.3(2) [online]: http://ustr.gov/trade-agreements/free-trade-agreements/peru-tpa/final-text; United States–Panama Trade Promotion Agreement, 28 June 2007, not yet in force, Article 17.3(2) [online]: http://ustr.gov/trade-agreements/free-trade-agreements/panama-tpa/final-text; United States–Chile Free Trade Agreement, 6 June 2003, in force 1 January 2004, (2003) 42 ILM 1026, Article 19.2(2); United States–Australia Free Trade Agreement, 18 May 2004, in force 1 January 2005, (2004) 43 ILM 1248, Article 19.2(2).

2 See, e.g. US–Colombia FTA, Articles 18.3(1), 18.12; DR-CAFTA, Articles 17.2(1), 17.10; US–Peru TPA, Articles 18.3(1), 18.12; US–Panama TPA, Articles 17.3(1), 17.11; US–Chile FTA, Articles 19.2(1), 19.6; US–Australia FTA, Articles 19.2(1), 19.7.

References

Beddoe, R., R. Costanza, J. Farley, E. Garza, J. Kent, I. Kubiszewski, L. Martinez, T. McCowen, K. Murphy, N. Myers, Z. Ogden, K. Stapleton and J. Woodward. 2009. Overcoming systemic roadblocks to sustainability: the evolutionary redesign of worldviews, institutions, and technologies. *Proceedings of the National Academy of Science* 106(8): 2483–2489.

Blais, L.E., and W.E. Wagner. 2008. Emerging science, adaptive regulation, and the problem of rulemaking ruts. *Texas Law Review* 86: 1701–1739.

Block, G. 2003. Trade and environment in the western hemisphere: expanding the North American Agreement on Environmental Cooperation into the Americas. *Environmental Law* 33(3): 501–546.

Bringezu, S., M. O'Brien and H. Schütz. 2012. Beyond biofuels: assessing global land use for domestic consumption of biomass: a conceptual and empirical contribution to sustainable management of global resources. *Land Use Policy* 29: 224–232.

Brown, P.G. 2015. Ethics for Economics in the Anthropocene, Chapter 2 in P.G. Brown and P. Timmerman, eds., *Ecological Economics for the Anthropocene*. New York: Columbia University Press.

Brown, P.G., and G. Garver. 2009. *Right Relationship: Building a Whole Earth Economy*. San Francisco: Berrett-Koehler.

Carpenter, S.R., and E.M. Bennett. 2011. Reconsideration of the planetary boundary for phosphorus. *Environmental Research Letters* 6: 1–12.

Clark, B., and J.B. Foster. 2009. Ecological imperialism and the global metabolic rift: unequal exchange and the guano/nitrates trade. *International Journal of Comparative Sociology* 50(30–34): 311–334.

Commission for Environmental Cooperation. 1995. Council resolution 95–07.

Commission for Environmental Cooperation. 1997. *Draft North American Agreement on Transboundary Environmental Impact Assessment* [online]: http://cec.org/Storage/92/8900_Draft-TEIA_Agreement-Oct97_en.doc.

Commission for Environmental Cooperation. 1999. *Council communiqué* [online]: http://cec.org/Page.asp?PageID=122&ContentID=1292.

Commission for Environmental Cooperation. 2001. *CEC Council communiqué* [online]: http://cec.org/Page.asp?PageID=122&ContentID=1290.

Commission for Environmental Cooperation. 2002. Letter from Alternative Representative for Canada Norine Smith to 2002 JPAC Chair Jonathan Plaut (14 June 2002) [online]: http://cec.org/Storage/25/1591_Council-Lessons-learned-04.pdf.

Commission for Environmental Cooperation. 2003a. *DDT no longer used in North America* [online]: http://cec.org/Storage/50/4285_DDT_en.pdf.

Commission for Environmental Cooperation. 2003b. Letter from Judith Ayres, Assistant Administrator, Office of International Affairs, United States Environmental Protection Agency, to JPAC Chair Gustavo Alanis-Ortega (24 March 2003) [online]: http://cec.org/Storage/24/1583_Counci-Response-02-12_en.pdf.

Commission for Environmental Cooperation. 2012. *CEC Ministerial Statement* [online]: http://cec.org/Page.asp?PageID=122&ContentID=25241.

Daly, H.E., and J. Farley. 2004. *Ecological Economics: Principles and Applications.* Washington, DC: Island Press.

Daniels, B. 2007. Emerging commons and tragic institutions. *Environmental Law* 37(3): 515–571.

Dobbs, R., J. Oppenheim, F. Thompson, M. Brinkman and M. Zornes. 2011. *Resource Revolution: Meeting the World's Energy, Materials, Food, and Water Needs.* New York: McKinsey Global Institute.

Ellis, E.C., J.O. Kaplan, D.Q. Fuller, S. Vavrus, K.K. Goldwijk and P.H. Verburg. 2013. Used planet: a global history. *Proceedings of the National Academy of Science* 110(20): 7978–7985.

Erb, K-H., F. Krausmann, V. Gaube, S. Gingrich, A. Bondeau, M. Fischer-Kowalski and H. Haberl. 2009a. Analyzing the global human appropriation of net primary production—processes, trajectories, implications. An introduction. *Ecological Economics* 69: 250–259.

Erb, K-H., F. Krausmann, W. Lucht and H. Haberl. 2009b. Embodied HANPP: mapping the spatial disconnect between global biomass production and consumption. *Ecological Economics* 69: 328–334.

Flores-Verdugo, F., P. Moreno Casasola, C.M. Agraz Hernandez, H. López Rosas, D. Benítez Pardo and A.C. Travieso Bello. 2007. La Topografía y el hidroperíodo: dos factores que condicionan la restauración de los humedales costeros. *Boletín de la Sociedad Botánica de México* 80(Suplemento): 33–47.

Foley, J.A., C. Monfreda, N. Ramankutty and D. Zaks. 2007. Our share of the planetary pie. *Proceedings of the National Academy of Science* 104(31): 12585–12586.

Foley, J.A., N. Ramankutty, K.A. Brauman, E.S. Cassidy, J.S. Gerber, M. Johnston, N.D. Mueller, C. O'Connell, D.K. Ray, P.C. West, C. Balzer, E.M. Bennett, S.R. Carpenter, J. Hill, C. Monfreda, S. Polasky, J. Rockström, J. Sheehan, S. Siebert, D. Tilman and D.P.M. Zaks. 2011. Solutions for a cultivated planet. *Nature* 478: 337–342.

G20. 2014. *G20 Leaders' Communiqué, Brisbane Summit* [online]: https://g20.org/wp-content/uploads/2014/12/brisbane_g20_leaders_summit_communique1.pdf

Gaines, S.E. 2003. Protecting Investors, Protecting the Environment: The Unexpected Story of NAFTA, Chapter 11 in D.L. Markell and J.H. Know, eds., *Greening NAFTA: The North American Commission for Environmental Cooperation.* Stanford CA: Stanford University Press.

Garver, G. 2008. Tooth decay. *The Environmental Forum* 25(3): 34–39.

Garver, G. 2013. The rule of ecological law: the legal complement to degrowth economics. *Sustainability* 5(1): 316–337 [online]: www.mdpi.com/2071-1050/5/1/316/htm

Garver, G. 2015. Forgotten Promises: Neglected Environmental Provisions of the NAFTA and the North American Agreement on Environmental Cooperation, book chapter in H. Kong and K. Wroth, eds., *NAFTA and Sustainable Development.* New York: Cambridge University Press.

Garver, G. 2019. Confronting remote ownership problems with ecological law. *Vermont Law Review* 43(3): 425–454.

Garver, G., and A. Podhora. 2008. Transboundary environmental impact assessment as part of the North American Agreement on Environmental Cooperation. *Impact Assessment and Project Appraisal* 26(4): 253–263.

Giljum, S., F. Hinterberger, C. Lutz and B. Meyer. 2009. Accounting and Modelling Global Resource Use, Chapter 8 in S. Suh (ed.), *Handbook of Input-Output Economics in Industrial Ecology*. PLACE: Springer Netherlands.

Gonzalez de Molina, M., and V.M. Toledo. 2014. *The Social Metabolism: A Socio-Ecological Theory of Historical Change*. New York: Springer.

Government of Canada. 2010. Canada lists emissions target under the Copenhagen accord [online]: http://ec.gc.ca/default.asp?lang=En&news=EAF552A3-D287-4AC0-ACB8-A6FEA697ACD6

Haberl, H., K-H. Erb, F. Krausmann, V. Gaube, A. Bondeau, C. Plutzar, S. Gingrich, W. Lucht and M. Fischer-Kowalski. 2007. Quantifying and mapping the human appropriation of net primary production in earth's terrestrial ecosystems. *Proceedings of the National Academy of Science* 104(31): 12942–12947.

Haberl, H., N.B. Schulz, C. Plutzar, K.-H. Erb, F. Krausmann, W. Loibl, D. Moser, N. Sauberer, H. Weisz, H.G. Zechmeister and P. Zulka. 2004. Human appropriation of net primary production and species diversity in agricultural landscapes. *Agriculture, Ecosystems and Environment* 102: 213–218.

Hails, C., S. Humphrey, J. Loh and S. Goldfinger, eds. 2008. *The Living Planet Report*. Gland, Switzerland: WWF International.

Häyhä, T., S.E. Cornell, H. Hoff, P. Lucas and D. van Vuuren. 2018. *Operationalizing the concept of a safe operating space at the EU level—first steps and explorations*. Stockholm Resilience Centre Technical Report, prepared in collaboration with Stockholm Environment Institute (SEI) and PBL Netherlands Environmental Assessment Agency. Sweden: Stockholm Resilience Centre, Stockholm University.

Häyhä, T., P.L. Lucas, D.P. van Vuuren, S.E. Cornell and H. Hoff. 2016. From planetary boundaries to national fair shares of the global safe operating space—how can the scales be bridged? *Global Environmental Change* 40: 60–72.

Heller, M.A. 1998. The tragedy of the anticommons: property in the transition from Marx to markets. *Harvard Law Review* 111(3): 621–688.

Hoffman, R.C. 2001. Frontier foods for late medieval consumers: culture, economy, ecology. *Environment and History* 7: 131–167.

Imhoff, M.L., L. Bounoua, T. Ricketts, C. Loucks, R. Harriss and W.T. Lawrence. 2004. Global patterns in human consumption of net primary production. *Nature* 429: 870–873.

Johnson, P.-M., and A. Beaulieu. 1996. *The Environment and NAFTA: Understanding and Implementing the New Continental Law*. Washington, DC: Island Press.

Johnson, P.-M, R. Page, J.A. Haverkamp, J.F. Mizroch, D. Basurto and B. Torres. 2004. *Ten Years of North American Environmental Cooperation: Report of the Ten-year Review and Assessment Committee to the Council of the Commission for Environmental Cooperation* [online]: http://cec.org/Storage/79/7287_TRACReport2004_en.pdf.

Jones, E.L. 2003. *The European Miracle: Environments, Economies and Geopolitics in the History of Europe and Asia, Third Edition*. New York: Cambridge University Press.

Knox, J.H., and D.L. Markell. 2012. Evaluating citizen petition procedures: lessons from an analysis of the NAFTA Environmental Commission. *Texas International Law Journal* 47(3): 505–540.

Krausmann, F., S. Gingrich, N. Eisenmenger, K.H. Erb, H. Haberl and M. Fischer-Kowalski. 2009. Growth in global materials use, GDP and population during the twentieth century. *Ecological Economics* 68: 2696–2705.

Lewis, S.L., and M.A. Maslin. 2015. Defining the Anthropocene. *Nature* 519: 171–180.

Lindgren, R.D. 2010. Letter from Richard D. Lindgren, Counsel, Canadian Environmental Law Association, to James Rajotte, Chair, Standing Committee on Finance, House of Commons (13 May 2010) [online]: http://cela.ca/publications/re-bill-c-9-proposed-changes-canadian-environmental-assessment-act.

M'Gonigle, M. 2008. Green legal theory: a new approach to the concept of environmental law. *Neue Konzepte* 4: 34–38.

Markell, D. 2010. The role of spotlighting procedures in promoting citizen participation, transparency and accountability. *Wake Forest Law Review* 45(2): 425–467.

McKinney, J.A. 2000. *Created from NAFTA: The Structure, Function, and Significance of the Treaty's Related Institutions*. Armonk, NY: M.E. Sharpe.

Meadows, D. 1999. *Leverage Points: Places to Intervene in a System.* Hartland VT: The Sustainability Institute.

Meadows, D., J. Randers and D. Meadows. 2004. *Limits to Growth: The 30-Year Update*. White River Junction, VT: Chelsea Green Publishing Company.

Moreno, I.S., J.W. Rubin, R.F. Smith III and T. Yang. 1999. Free trade and the environment: the NAFTA, the NAAEC, and implications for the future. *Tulane Environmental Law Journal* 12: 405–478.

Nordhaus, W. 2007. *The Challenge of Global Warming: Economic Models and Environmental Policy.* New Haven, CT: William Nordhaus.

Nordhaus, W.D. and Z. Yang. 1996. A regional dynamic general-equilibrium model of alternative climate-change strategies. *The American Economic Review* 86(4): 741–765.

Nyquist, B., A. Persson, F. Moberg, L. Persson, S. Cornell and J. Rockström. 2013. *National Environmental Performance on Planetary Boundaries: A Study for the Swedish Environmental Protection Agency.* Stockholm: Swedish Environmental Protection Agency.

Orr, D.W. 2011. *Hope is an Imperative.* Washington: Island Press.

Parent, M. 2007. Letter from the Nova Scotia Minister of Environment & Labor to P.G. Buxton, Bilcon of Nova Scotia Corporation [online]: www.novascotia.ca/nse/ea/whitespointquarry/WhitesPointQuarryMinDecision.pdf

Pearce, F. 2012. *The Land Grabbers: The New Fight Over Who Owns the Earth.* Boston: Beacon Press.

Pelletier, N. 2010. Of laws and limits: an ecological economic perspective on redressing the failure of contemporary global environmental governance. *Global Environmental Change* 20: 220–228.

Polanyi, K. 1957. *The Great Transformation: The Political and Economic Origins of Our Time*. Boston: Beacon Press.

Pollard, D., R. Almond, E. Duncan, M. Grooten, L. Hadeed, B. Jeffries and R. McLellan, eds. 2010. *Living Planet Report 2010*. Gland, Switzerland: World Wide Fund for Nature.

Pomeranz, K. 2002. Political economy and ecology on the eve of industrialization: Europe, China, and the global conjuncture. *The American Historical Review* 17(2): 425–446.

Rees, W.E. 2002. Globalization and sustainability: conflict or convergence? *Bulletin of Science, Technology & Society* 22.

Rockström, J., W. Steffen, K. Noone, Å. Persson, F.S. Chapin III, E. Lambin, T.M. Lenton, M. Scheffer, C. Folke, H. Schellnhuber, B. Nykvist, C.A. De Wit, T. Hughes, S. van der Leeuw, H. Rodhe, S. Sörlin, P.K. Snyder, R. Costanza, U. Svedin, M. Falkenmark, L. Karlberg, R.W. Corell, V.J. Fabry, J. Hansen, B. Walker, D. Liverman,

K. Richardson, P. Crutzen and J. Foley. 2009. Planetary boundaries: exploring the safe operating space for humanity. *Ecology and Society* 14(2): 32 [online] URL: www.ecologyandsociety.org/vol.14/iss2/art32/
Sachs, J.D. 2008. *Common Wealth: Economics for a Small Planet.* New York: The Penguin Press.
Samet, J.M., and T.A. Burke. 2020. Deregulation and the assault on science and the environment. *Annual Review of Public Health* 41: 347–361.
Security and Prosperity Partnership. 2005. U.S. Department of State Archives [online]: http://2001-2009.state.gov/p/wha/rt/spp/
Speth, J.G. 2008. *The Bridge at the Edge of the World.* New Haven, CT: Yale University Press.
Srinivasan, U.T., S.P. Carey, E. Hallstein, P.A.T. Higgins, A.C. Kerr, L.E. Koteen, A.B. Smith, R. Watson, J. Harte and R.B. Norgaard. 2008. The debt of nations and the distribution of ecological impacts from human activities. *Proceedings of the National Academy of Science* 105(5): 1768–1773.
Steffen, W., K. Richardson, J. Rockström, S.E. Cornell, I. Fetzer, E.M. Bennett, R. Biggs, S.R. Carpenter, W. de Vries, C.A. de Wit, C. Folke, D. Gerten, J. Heinke, G.M. Mace, L.M. Persson, V. Ramanathan, B. Reyers and S. Sörlin. 2015. Planetary boundaries: guiding human development on a changing planet. *Science* 347(6223): 1259885-1–1259885-10.
United Nations. 2012. *The Future We Want.* New York: United Nations [online]: https://documents-dds-ny.un.org/doc/UNDOC/GEN/N11/476/10/PDF/N1147610.pdf?OpenElement.
United Nations. 2015a. *Transforming Our World: The 2030 Agenda for Sustainable Development.* New York: United Nations.
United Nations Environment Programme (UNEP). 2011. *Towards a Green Economy: Pathways to Sustainable Development and Poverty Eradication—A Synthesis for Policymakers.* Nairobi: UNEP. www.unep.org/greeneconomy/GreenEconomyReport/tabid/29846/language/en-US/Default.aspx.
Vaughan, S. 2004. Thinking North American Environmental Management, Chapter 5 in T.J. Courchene, D.J. Savoie and D. Schwanen, eds., *The Art of the State II: Thinking North America.* Montreal: Institute for Research on Public Policy.
Victor, P., and T. Jackson. 2012. *Towards an Ecological Macroeconomics.* Paper prepared for the INET Annual Plenary Conference. Berlin: INET [online]: http://inet economics.org/uploads/papers/victor-peter-berlin-paper-draft-w-Tim-Jackson.pdf.
Vitousek, P.M., P.R. Ehrlich, A.H. Ehrlich and P.A. Matson, 1986. Human appropriation of the products of photosynthesis. *BioScience* 36(6): 368–373.
Vizcaino, J. 2012. Stephen Harper sells resource-friendly economy to Latin American execs. *Toronto Star*, 12 April 2012 [online]: http://thestar.com/business/2012/04/14/stephenharpersellsresourcefriendlyeconomytolati americanexecs.html.
Watson, J.K.R. 2013. *The WTO and the Environment: Development of Competence Beyond Trade.* New York: Routledge.
Wold, C. 2010. Taking stock: trade's environmental scorecard after twenty years of "trade and environment." *Wake Forest Law Rev*iew 45(2): 319–354.
Worster, D. 1987. The vulnerable Earth: toward a planetary history. *Environmental Review* 11(2): 87–103.
Wright, D.H. 1990. Human impacts on energy flow through natural ecosystems, and implications for species endangerment. *Ambio* 19(4): 189–194.
WTO. 2020. World Trade Organization website [online]: www.wto.org.

Young, O., F. Berkhout, G.C. Gallopin, M.A. Janssen, E. Ostrom and S. van der Leeuw. 2006. The globalization of socio-ecological systems: an agenda for scientific research. *Global Environmental Change* 16: 304–316.

Zepeda, E., T.A. Wise and K.P. Gallagher. 2009. *Rethinking trade policy for development: lessons from Mexico under NAFTA*. Carnegie Endowment for International Peace [online]: http://carnegieendowment.org/publications/index.cfm?fa=view&id=24271.

Part V

Degrowth and ecological law

15 The degrowth movement as a testing ground for ecological law

Significant obstacles stand in the way of a transition to a dominant global legal system that promotes a mutually enhancing human-Earth relationship. The mostly theoretical notion of law that I present in this book is a future vision not only of law, but also of the social, cultural, political and economic contexts in which law is embedded (M'Gonigle 2008). It is a vision that implies an inevitable evolution away from current political orders and power structures, toward new ones that must emerge as this transition unfolds. The qualitative analysis in this book shows that to attain this vision, the resistance to change in the legal system and related systems is high and will take a long time to overcome. This is particularly true with respect to necessary paradigm shifts—especially the transition from the dominant anthropocentric growth-insistent paradigm to the human-inclusive ecocentric limits-insistent paradigm I advocate. This transition implies a significant change in deeply entrenched power structures and political orders at all scales—change that will likely emerge in unpredictable ways and through unpredictable actors (Brown and Garver 2009).

I propose in this book an analytical framework for rigorously exposing obstacles to and opportunities for change in view of lock-in tendencies in law and related systems and then developing strategies and priorities for overcoming the resistance that is revealed. The framework is intended to have the flexibility to be useful to a wide range of potential agents in the proposed transition—from informed individuals and engaged activists seeking social change with bottom-up approaches to policy and decision-makers in more formal settings and power structures. Currently, nearly all those power-laden policy and decision-makers are likely resistant to this transition. Thus, the broad and coordinated research that would be best suited to developing the strategies and priorities in this transition will have to start elsewhere. The degrowth movement is a prime candidate.

Sustainable development has shown little signs of being capable of triggering the "overarching policies and radical change of behaviour needed at individual and collective scales" (Martinez-Alier et al. 2011) needed to curb the rampant consumerism that drives patterns of social metabolism that are undermining the capacity of the Earth's ecosystems to support human and other life. The degrowth movement is an emerging alternative response to humanity living beyond its means. Degrowth (or "sustainable degrowth") is "a downscaling of production and consumption that increases human well-being and enhances ecological conditions and equity on the planet" (Research & Degrowth 2011). It is a "ruthless critique of the dogma of more—the dogma of economic growth" (Kallis 2018: vii). Indeed, within heterodox economics, degrowth is the strongest, most direct critique of economic growth (D'Alisa et al. 2015; Kallis 2018). Degrowth is therefore a forceful challenge to the growth-insistent sustainable development model (Martinez-Alier et al. 2010; Kallis 2011), and a more hopeful approach to long-term perpetuation of a mutually enhancing human-Earth relationship. Because its core ideology resonates strongly with the ideas in this book, and because it intentionally combines academic research, activism and support for real-world experiments and experiences that give life to degrowth ideas, the degrowth movement is a natural testing ground for ecological law and the research approaches presented in this book.

Degrowth has roots both in the quest of ecological economics for a steady-state economy that respects ecological limits and in the socio-cultural objections to wealth accumulation and bigness reflected in Schumacher's *Small is Beautiful* and the earlier work of French intellectuals who critiqued "gigantism" (Martinez-Alier et al. 2010). Although the term *décroissance* can be traced back at least to the 1970s (Martinez-Alier et al. 2010; D'Alisa et al. 2015), the current degrowth movement took root with the publication of special issues on *décroissance* of the French magazine *Silence* in February and March 2002 (D'Alisa et al. 2015). The contemporary degrowth movement includes a collaborative discussion and research agenda aimed at developing "a framework for transformation to a lower and sustainable level of production and consumption" (Research & Degrowth 2011).

By suggesting that the human enterprise is too big by some measure, such that it must downsize, degrowth implicitly suggests a limit on the size of the global economy. Indeed, limits are part of the essence of degrowth. Yet, degrowth goes beyond ecological economics, sustainability and other expressions of the idea that humans must live within the Earth's means—and therefore that an infinitely growing economy is biophysically impossible—by explicitly incorporating the notion that the economy has already exceeded those limits. The unmistakable warning of the jolting term *degrowth*, which derives from the French word *décroissance*, is not that the economy can and might become too big, but rather that it already is too big and therefore must degrow.

As a strong reaction to growth-insistent perspectives, degrowth takes on a negative connotation that has been criticized (Kallis 2018: 158). In one sense, the jarring impact of the term "degrowth" is appropriately aimed at shaking loose the human imagination from the overwhelmingly dominant idea that the economy must grow for humanity to survive. Although good reasons currently exist for degrowth to speak the language of the system that it seeks to undermine (Kallis 2011), growth in fact causes many things to decline. In view of the future that the degrowth movement is working toward, degrowth should also be flipped around so as to promote the positive growth of those things that decline as the economy grows: ecological integrity, biodiversity, cultural diversity, compassion, equity, respect for life, human solidarity, simplicity, vegetarianism—and ultimately, the prospect for all beings to live a good life.

Degrowth combines a commitment to respecting ecologically based limits with a commitment to developing a comprehensive, practicable approach to building thriving human communities based on conviviality and human solidarity without consumerism or material and energy excess. The reforms associated with degrowth "emphasize redistribution (of work and leisure, natural resources and wealth), social security and gradual decentralization and relocalisation of the economy, as a way to reduce throughput and manage a stable adaption to a smaller economy" (Kallis 2011: 876).

The full meaning of degrowth and the objectives of the degrowth movement remain fluid (Kallis 2018). What limits have been exceeded, and what needs to degrow? To be consistent with a mutually enhancing human-Earth relationship, the overriding limits inherent in degrowth are the ecological limits within which human society must operate in order to flourish in successive generations, while also allowing the flourishing of other components of Earth systems. Other limits may be relevant as well—on the size or scope of enterprises or institutions, on capital accumulation, on incomes and individual and collective monetary wealth, on consumption, on work time, on technological innovations (D'Alisa et al. 2014; Kallis 2018)). However, from the systems-based perspective on which ecological economics is founded, the paramount normative boundaries that should contain the collective activities of humans are ecological and are tied to the throughput of material and energy in the economy and the ecological impacts of that throughput (Beddoe et al. 2009; Pelletier 2010).

In regard to how to deal collectively with the aggregated pressures on global ecological limits that derive from geographically and temporally diverse sources, the question of the relationship between localities horizontally with each other and vertically with other levels of political order is particularly cogent. In general, degrowth thinkers and activists tend to be skeptical of the possibility of true democracy at higher levels of political organization, on the grounds that "[d]emocracy can probably only function where the polis is small and firmly anchored to a set of values" (Latouche 2006). Hence the proposition that

> [t]he relationships between the polities within the global village could be regulated by a democracy of cultures, in what might be called a pluriversalist vision. This would not be a world government, but merely an instance of minimal arbitration between sovereign polities with highly divergent systems.
>
> (Latouche 2006)

Seeds of this kind of democracy may be taking root in the many eco-villages, eco-communes, cooperatives and other collective bodies at the local to national scale that use methods of direct participatory democracy for decision-making (Kallis 2018: 119, 135).

The dynamics and interregional interdependencies of biogeochemical, geologic, hydrologic, climatic, atmospheric and other processes prevent any sub-global region or locality from being isolated from the rest of the integrated global ecosystem (Häyhä et al. 2016). A study that detected fingerprinted dioxins from sources in Mexico, the United States and southern Canada in mothers' milk and other receptors in Inuit communities in Nunavut illustrated this interdependency well (Commoner et al. 2000). How does a locality protect itself from impacts that arise outside of it, and what ensures that it will be responsible for impacts it causes that transcend its borders? A central challenge for degrowth, which degrowth researchers and activists are already taking on, is to develop legal and policy mechanisms for fairly distributing the responsibility to respect global or regional ecological limits down to the local level. This architecture of distribution should also incorporate mechanisms for enabling all humans and other living beings to flourish, built on principles of intragenerational, intergenerational and interspecies fairness (Bosselmann 2008; Brown and Garver 2009).

The European principle of subsidiarity, discussed in Chapter 11, provides a way to reconcile a preference for establishing policy at the local level with the reality that localities are never immune from impacts caused by activities in other localities nor able to avoid causing impacts that reach other localities. The challenge in applying the subsidiarity principle is to account for a broad range of cultural, ecological and socio-political contexts, and a plurality in the way behavior is made to conform to limits, in fashioning a predictable and consistent global system for implementing limits such as planetary boundaries. Subsidiarity should be implemented in recognition that global governance includes the participation of "a long list of institutions including governments, businesses, non-governmental organizations (NGOs), universities, research centers, and foundations [operating] inside and outside of government and across national and institutional boundaries" (Saunier and Meganck 2009: 3–4). Although degrowth calls for significant reforms in these institutions, the principle of subsidiarity should nonetheless be useful for allocating governances roles among political orders at different scales.

The organization Research & Degrowth,[1] established in 2006, has been maintaining the focus of the degrowth movement by spearheading and overseeing a series of degrowth conferences and helping to maintain a repository of information about degrowth that has emerged from conferences and other venues.[2] A key outcome of the Barcelona degrowth conference in 2010 was a more comprehensive outline of a core set of degrowth proposals developed during the Barcelona gathering and a previous conference in Paris (Barcelona Conference 2010), and those ideas continue to evolve. The degrowth concepts in the Barcelona "bullet points" are organized around Democracy, Education, Social Economy, Natural Resources and Demography. The bullet points emphasize community involvement, decommodification and decommercialization with respect to all of these categories, with a focus on reorienting the economy toward local autonomy, equitable sharing, low-impact technologies, a more restricted view of private property, food sovereignty and floors and ceilings on income, as well as on monetary reform, trade reform, constraints on advertising and moratoria on harmful technologies. The overarching proposal with regard to natural resources is to reduce and then maintain the throughput of energy and materials within the Earth's life support capacity.

The outcomes of the Barcelona degrowth conference have been built upon in subsequent international conferences in Vienna (2012), Leipzig (2014), Budapest (2016) and Malmö (2018), as well as several regional and local conferences. The research areas identified across these gatherings raise questions about the role of law and of more informal rules of conduct based on culture and tradition that legal pluralism encompasses (M'Gonigle 2008; Garver 2013). The discursive and evolving process to give meaning to degrowth has continued since Barcelona, especially in the Group Assembly Process (GAP) at the Leipzig degrowth conference in 2014, a process that continued on the web after the conference (Leipzig Degrowth 2014). The analytical approach I propose in this book, with a focus on lock-in assessments, holds considerable promise to help move degrowth research and activism forward.

Distilling this ongoing refinement of degrowth ideas, Kallis (2018: 118–122) proposes nine core principles of the vision of degrowth: (1) End to exploitation, (2) Direct democracy, (3) Localized production, (4) Sharing and the commons, (5) Provision of relational goods, (6) Unproductive expenditures geared to communal activities, (7) Care, (8) Diversity and (9) Decommodification. These translate into a "degrowth 'policy package'" (Kallis 2018: 127):

- GDP being abolished and substituted with other indicators of human and ecological well-being;
- Work being shared, with a reduction of working hours to create employment in the absence of growth;
- A universal income or guaranteed bundle of public services, ensuring that everyone has enough to get by without depending on money;
- Redistributive taxation to increase equality and the establishment of maximum income to arrest competition for positional consumption;
- A redirection of public investments from the private sector to the public, and from infrastructure and activities that increase productivity to expenditures that green the economy and reclaim the commons; and
- Environmental limits and taxes to finance low-income groups.

These principles and policies, along with others that the degrowth movement has generated and continues to generate, are potential topics for the lock-in/lock-out assessments described in this book. In the area of law and degrowth, I have developed a methodology for ecological law case studies that can make use of lock-in/lock-out assessments, Sbert's lens of ecological law (Sbert 2020) and other analytical tools and research methods that case study teams might choose. Ecological law case studies involve assessment of real-world situations or activities (e.g. extraction projects, international trade in controversial products such as palm oil, development policies) according to principles of ecological law. The main objectives of these case studies is first, to give more concreteness to how ecological law is not just about transforming environmental law, but all law; and second, to show how ecological law is not just theoretical, but can be imagined and applied in real-world situations. Study teams first deconstruct the selected projects, policies, sectors or other proposals or activities to reveal all the legal and policy realms that are involved, taking a broad and open approach. Relevant legal and policy realms could include any of the elements in Table 9.1 in Chapter 9. The team them analyzes the law and policy supporting the proposal or activity, using tools like the lens of ecological law. Finally, the team reconstructs the proposal or activity using principles of ecological law—or possibly decides that in a regime of ecological law, the proposal or activity should not go forward. Lock-in/lock-out assessments would be most relevant at this final stage. For example, a mining project could be deconstructed, analyzed and reconstructed around key local, State/provincial, national and international laws, regulations and policies that apply to every aspect of the project, such as those related to ownership regimes and organizational structures for all of the entities involved, labor laws for workers, health care for the community, other provisioning for the community and so on.

Other social movements capture this positive vision of change toward human solidarity, community and peaceful relations, respectful of the limits of Earth's life support capacity and the intergenerational needs of all living beings. Two examples among many others bear mention. The Transition Town movement engages communities at the local scale to work together to develop and implement plans for long-term resilience in the face of climate change and other environmental challenges (Hopkins 2008). In South America, an emerging social movement is coalescing around the notion of *buen vivir. Buen vivir* emphasizes well-being based on place-based notions of communal solidarity that includes the full community of life, as well as recognition and preservation of diversity of knowledge and ways of knowing (Gudynas 2015). These and other social movements aimed at radical transformation of the human-Earth relationship—in North America, the Occupy and Idle No More movements are notable—have implications for law and are prime arenas for the participatory research proposed here.

The degrowth movement, Transition Towns and social movements built around *buen vivir* are merely illustrative and not meant to limit the contexts in which the conceptual approach proposed in this book might be refined and adapted in a wide range of contexts. By highlighting an integrative framework for considering the role of law in the necessary transformation, and for focusing on strategies and priorities for overcoming resistance to change, my aim is to contribute to innumerable positive future visions for the human-Earth relationship in an active way. One desired outcome in proposing this framework and the diverse transdisciplinary research that is possible under it is to make the need for transformative change more concrete and visible. The framework is intended to be applicable to research and science-based activism and practice to develop priorities and strategies for embedding mutual human-Earth enhancement at all scales, local to global, and across them, and in a wide range of subject areas. Moving forward, the approaches I present in this book hold promise to help ignite many sparks with potential to flare up and light the way to a more positive path for people and Earth.

Notes

1 Research & Degrowth (R&D) [online]: https://degrowth.org.
2 This repository of degrowth information is currently at www.degrowth.info/en/.

References

Barcelona Conference on Economic Degrowth for Ecological Sustainability and Social Equity [Barcelona Conference]. 2010. *"Degrowth bullet points" from the Barcelona conference.*

Beddoe, R., R. Costanza, J. Farley, E. Garza, J. Kent, I. Kubiszewski, L. Martinez, T. McCowen, K. Murphy, N. Myers, Z. Ogden, K. Stapleton and J. Woodward. 2009. Overcoming systemic roadblocks to sustainability: the evolutionary redesign of worldviews, institutions, and technologies. *Proceedings of the National Academy of Science* 106(8): 2483–2489.

Bosselmann, K. 2008. *The Principle of Sustainability: Transforming Law and Governance.* Burlington VT: Ashgate Publishing Co.

Brown, P.G., and G. Garver. 2009. *Right Relationship: Building a Whole Earth Economy.* San Francisco: Berrett-Koehler.

Commoner, B., P.W. Woods, H. Eisl and K. Couchot. 2000. *Long-range Air Transport of Dioxin from North American Sources to Ecologically Vulnerable Receptors in Nunavut, Arctic Canada: Final Report to the North American Commission for Environmental Cooperation.* Montreal: Commission for Environmental Cooperation.

D'Alisa, G., F. Demaria and G. Kallis, eds. 2014. *Degrowth: A Vocabulary for a New Era.* New York: Routledge.

Garver, G. 2013. The rule of ecological law: the legal complement to degrowth economics. *Sustainability* 5(1): 316–337 [online]: www.mdpi.com/2071-1050/5/1/316/htm

Gudynas, E. 2015. Buen Vivir, Chapter 48 in G. D'Alisa, F. Demaria and G. Kallis, eds., *Degrowth: A Vocabulary for a New Era.* New York: Routledge.

Häyhä, T., P.L. Lucas, D.P. van Vuuren, S.E. Cornell and H. Hoff. 2016. From planetary boundaries to national fair shares of the global safe operating space—how can the scales be bridged? *Global Environmental Change* 40: 60–72.

Hopkins, R. 2008. *The Transition Handbook: From Oil Dependency to Local Resilience.* White River Junction VT: Chelsea Green Publishing.

Kallis, G. 2011. In defence of degrowth. *Ecological Economics* 70: 873–880.

Kallis, G. 2018. *Degrowth.* Newcastle upon Tyne, UK: Agenda Publishing.

Latouche, S. 2006. The globe downshifted. *Le Monde Diplomatique* (English edition), 13 January 2006 [online]: mondediplo.com/2006/01/13degrowth.

Leipzig Degrowth. 2014. Leipzig 2014 [online]: www.degrowth.de/en/leipzig-2014/.

M'Gonigle, M. 2008. Green legal theory: a new approach to the concept of environmental law. *Neue Konzepte* 4: 34–38.

Martinez-Alier, J., H. Healy, L. Temper, M. Walter, B. Rodriguez-Labajos, J-F. Gerber and M. Conde. 2011. Between science and activism: learning and teaching ecological economics with environmental justice organisations. *Local Environment: The International Journal of Justice and Sustainability* 16(1): 17–36.

Martinez-Alier, J., U. Pascual, F-D. Vivien and E. Zaccia. 2010. Sustainable de-growth: mapping the context, criticisms and future prospects of an emergent paradigm. *Ecological Economics* 69: 1741–1747.

Pelletier, N. 2010. Of laws and limits: an ecological economic perspective on redressing the failure of contemporary global environmental governance. *Global Environmental Change* 20: 220–228.

Research & Degrowth. 2011. Definition [online]: www.degrowth.org/definition-2.

Saunier, R.E., and R.A. Meganck. 2009. *Dictionary and Introduction to Global Environmental Governance, Second Edition*. London: Earthscan.

Sbert, C. 2020. *The Lens of Ecological Law: A Look at Mining.* Cheltenham, UK: Edward Elgar Publishing.

Index

absolute advantage in trade 185–6
accumulation: capital 229; of impacts 14, 18; wealth 50, 228
acidification: acid rain 3, 51, 82–4; Acid Rain Program 73, 82–4; ecosystems 81; ocean 115, 117, 119
adaptive management 112, 177
adaptiveness: definition of, in systems 173; of Earth systems 142; in ecological law 4, 102, 128, 146–7, 156, 174, 177, 209–11; of institutions 203; and resilience 5, 169–71, 173; of societies 109, 110; *see also* complex adaptive systems
advertising 49, 50, 231
aerosol loading 115, 116, 124
affluence 7, 34, 35, 67, 123, 132; *see also* IPAT
Africa 22, 29, 30, 31, 37, 46, 50, 185
agency and agents 44, 109, 110, 137, 138, 142, 227
aggregate impacts of human activity 51, 52, 67, 187, 212
agriculture: agricultural run-off 75; agricultural surplus 17; and anthropogenic biomes 98, 163; automation in 32; and critical levels in the EU 80; ecological impacts of 30, 31, 32–3, 45, 116–17, 118; and emergence of social elite 31, 44; in environmental history 28, 29; extensification 31; and first written legal codes 44; and HANPP 206–7; and interregional problem shifting 118; and land and resource grabbing 32; modern industrial 32–3, 34; monoculture 154; Neolithic Transition 2, 3, 12, 29, 30–3; and nutrient cycles 116–17; and population growth 35–6, 44, 208; sector, as an arena for law 111; terraced, in ecological restoration 160; and US water quality standards 74; yields 36, 80
agricultural and agrarian societies 2–3, 12, 17, 29, 30–3
air pollution and air pollutants: and agriculture 31; control 73, 76–84, 132, 141, 195; and leverage points for systems change 172; long-range 80–4, 208; *see also* acidification
Amazon 118
ambient air and water quality 74, 77, 79–80, 81
America: the Americas 16, 29, 30, 31, 32, 37, 46, 49, 69, 183, 185; Central America 139; Latin America 50, 120, 196; North America 6, 54, 133–4, 178, 185, 188–204, 232; South America 232
American Clean Energy and Security Act of 2009 (ACES) 79
American Revolution 46
ammonia 81
analytical framework 122, 227
Anthropocene 2, 3, 6, 22, 23, 98, 108, 127, 153, 163, 170, 171, 177: Anthropocene Working Group (AWG) 2, 11, 13; conceptual factors leading to 42–57, 109; criticism of term 14–17; emergence of 28–9, 183; formal recognition of 11, 12–13; as frame of reference 11–22; Industrial Revolution as the start of 33; key challenges in 24, 65, 93, 154, 174; law in 2, 93, 153, 163, 174, 180, 205; narrative 14; Paleoanthropocene 13; trade in 204–5
anthropocentrism 65–7, 98, 99, 104, 127–8, 162, 172, 227
anthropogenic biomes 98, 100

appropriation, of human net primary productivity (HANPP): as metric of social metabolism 6, 122, 170, 206–8; relevance to ecological law 129, 131, 141
arbitration 54, 111, 139, 198, 201–2, 229
artificial intelligence *see* intelligence, artificial
Asia 22, 29, 30, 33, 37, 46, 50, 118, 185: East Asia 45, 185
assembly rules theory 159
atmosphere 34, 80, 82, 114, 115, 129, 133, 141, 175, 176
attachment to place 152, 161–4
Austria 122
authoritarianism 56

Bacon, F. 46
beavers 103
bees 66–7
behavior, self-interested 23
benchmarks of ecological integrity 153, 154, 157
Berry, T. 93–4, 97, 99–100, 101
biocapacity of Earth 120, 130
biofuels 36, 208
biogeochemical flows and processes 75, 81, 84, 115, 116, 118, 206, 208, 230
biomass: energy from 31; harvest of 4, 128, 147, 206–8; as net primary production in plants 206; in trade 206
biodiversity: as an arena for law 111; Biodiversity Intactness Index 117, 156; and degrowth 228; and ecological restoration 158, 160; effects of agriculture on 31, 32; and HANPP 206–8; and IPAT 35; loss of or change in 12, 24, 54, 67, 82, 127; and monetary valuation of ecosystems 66; and planetary boundaries 116, 117, 118, 129, 140, 145, 156; as a public good 56; and resilience 173; and trade 188, 200, 204, 217
bioproductive land as part of ecological footprint 131
bioprospecting 50
biosphere: integrity 114, 116, 117, 118, 153, 156, 157, 207; models 131
biotechnology 34
biotic community *see* community
birth rate 36
Black Death 46
Bolivia 103
Boserup, E. 36
Brazil Tire cases 188
bubbles, capital 49
buen vivir 6, 232
buildings: carbon dioxide emissions from 133; codes 178, 179; energy consumption of 133; green 70, 133; sector 34, 133–4
burden of proof 104, 144
Bush Administration 75

campaign finance 173
Canada: citizen submissions regarding 200; and dioxin pollution 118, 230; ecological footprint in 120; energy use in 134; environmental backsliding in 193–4, 196; and NAFTA Chapter 11, 54, 201–2; and trade 6, 139, 189, 190, 192, 197, 212
cancer 50
capabilities, bounded 119
cap-and-trade 172: acid rain 73, 82–4; greenhouse gases 79
capacity: adaptive 173, 175; carbon storage 31, 33; carrying 37, 123, 124; ecological 67, 68, 132, 133; intellectual 3, 16, 17, 30, 101, 163; governance 55, 136; life support, of Earth 1, 5, 6, 12, 31, 49, 57, 98, 119, 135, 136, 137, 169, 184, 216, 227, 231, 232; *see also* biocapacity
capital: accumulation 229; capitalism 16, 17, 37, 46–50, 51, 53, 54, 69, 173; Capitalocene 17; mobility of 50, 136, 186, 188; natural 47, 56, 66, 133
carbon capture and sequestration 79
carbon cycle 66
carbon dioxide: atmospheric concentration of 114, 115, 116, 117, 121, 175, 176; emissions 37, 133; and planetary boundaries 114, 115, 116, 121, 175, 176; regulation of 73, 76–80, 84
carbon monoxide 80
carbon tetrachloride 84
carrying capacity *see* capacity
Cartesian worldview 65, 98
catastrophe: catastrophic change or collapse 3, 14, 15, 17, 23, 25, 35, 51, 73, 79, 85, 114, 115, 116, 134, 137, 142, 146, 156, 175, 176, 177, 188; as feature of complex adaptive systems 12, 108; short-term 36, 61
caution *see* precaution and precautionary principle
Central America *see* America

centralization 56, 136
change: global 11, 12, 13, 15, 24, 137, 183; transformative 169; resistance to 169, 173, 227, 232
chaos and chaotic systems behavior 12, 115
cheap labor *see* labor
Chile 50
chilling effect on regulation 57, 70
China 31, 45, 47: Loess Plateau 160
chlorofluorocarbons [CFCs] 84–5
choice: collective 123, 157, 160; consumer 23; ethical 35; human 158; moral 14, 17; normative 95; political 116, 144, 163
Christianity 46, 56
citizen suits 140
citizen(s) 110, 111, 140
Citizens United v. *Federal Election Commission* 173, 174
civilization(s) 3, 14, 31, 34, 44, 45, 53, 109, 110
class and class structure 44, 45
Clayton/Bilcon v. *Government of Canada* 201
Clean Air Act, U.S. 73, 76–80, 82–4, 195
Clean Water Act, U.S. 73–6, 195
climate change: and agriculture 32–3; causes and drivers of 15, 16, 37, 121, 207; compared to stratospheric ozone depletion 85; and ecological law 5; and economic models 66; and EROEI 208; inadequate control of 3, 54, 190, 194; and lock-in/lock-out assessments 175–9, 215; and "merchants of doubt" 51; mitigation of 73, 76–80, 120, 146, 174; as planetary boundary 114, 117, 118, 119, 121, 156; and precautionary principle 144–6; relation to other planetary boundaries 117, 118; systemic nature of 3, 19, 144; and trade 203, 205, 215; and Transition Towns 232; trends 200; *see also* Paris Agreement
Clinton Administration 75, 189
Club of Rome 114
coal: and the Anthropocene 16; coal-fired power plants 51, 76, 78, 83; and the Industrial Revolution 34, 35; low-sulfur 83
Code of Hammurabi 44, 110
Code of Ur-Nammu 44
collapse: civilizational or societal 3, 17, 45, 109, 110, 121; of Soviet Union 47
collective action 51, 170
colonialism 16, 17, 32, 37, 46
colonization 28, 30, 45, 131
Columbian Exchange 32, 183
Columbus, C. 183
comfort 23
command-and-control 83, 84
commerce 28, 35, 111, 187; *see also* trade
Commission for Environmental Cooperation (CEC): budget 197; cooperative program of 189, 191; Council 191, 192, 195, 196, 197, 198; creation of 189, 191; factual records 189, 199, 200, 202; failures of 190, 193–200; Joint Public Advisory Committee 191, 197; Secretariat 191, 197, 198; submissions on enforcement matters 191, 199–200, 202; successes of 189; *Taking Stock* report 189
commodification and decommodification 102, 230, 231
common but differentiated responsibilities 138, 146
common good 49
common law *see* law
common pool resources (CPRs) 55, 157, 160
commons 58, 94, 99, 131, 137, 188, 204, 209, 231; *see also* Tragedy of the Commons
commonwealth of life 6, 23, 120, 137, 153, 184
communication(s) 30, 34, 122
communalism and communal goods 42, 44, 45, 159, 177, 231, 232
communion 99
community: -based research 217; biotic or biologic 52, 75, 129, 153; and degrowth 232; dialogue and engagement 157, 158, 230; Earth 53, 94, 99, 100–3, 130; ecology 159; European *see* Europe; global 162; human 44, 128, 142, 175; international 48, 85, 183, 186, 187, 211, 212; life 42; of life 2, 17, 53, 66, 97, 99, 100, 104, 128, 130, 137, 162, 232; local 161; scientific 13, 143; well-being 152
comparative advantage in trade 185–6, 188
compensation: just, for taking of property 56, 57, 70, 202, 211; monetary 145, 163
competition: competitive advantage 47; international 45, 54; market 50, 67, 186, 208
complexity 12, 30, 35, 109, 147, 177, 207; *see also* complex adaptive systems

complex adaptive systems: characteristics of 5, 109, 110, 112, 175; cyclical changes in 154, 155; generally 1, 108, 174; human decision-making as 109; law as 1, 2, 43, 108–12, 171, 177; *see also* panarchy
conceptual factors of the human-Earth dilemma 1, 2, 3, 11, 42–57
conflict resolution 160, 161
conquest 31, 37, 45, 46, 53
constitutional law 56, 70, 93, 104–5, 136, 172, 173, 178, 215
consumer culture and consumerism 47, 49, 132, 227, 229
consumers 56, 190, 204, 212
consumption: constraints on 134, 138, 211; and degrowth 228, 229, 231; and ecological footprint 32, 120, 185; and economic growth 137; energy 133, 134, 137; freshwater 115; and HANPP 207, 208; increase 48, 57, 131, 185; and IPAT 7, 35, 123–4; legally protected 67, 204; meat 32, 208; overconsumption 188, 211; of ozone-depleting substances 84, 85; paper 130; and planetary boundaries 138, 206; rate 36, 37; sustainable 152; unlimited 67
contracts 56, 111, 177, 187
Convention on Long-range Transboundary Air Pollution (LRTAP) 73, 80–2, 132, 208
cooperation 67, 192, 209
cooperatives 229
coral reefs 140
corporations 46, 47, 50, 54, 99, 111, 137, 173, 186
critical loads and levels 73, 80–2, 83, 86, 141, 208
cropland 33
cult of wilderness 98, 102, 153
culture(s) and cultural systems 24, 44, 45, 53, 100, 111, 130, 135, 158, 229, 231
cumulative impacts 52, 71
custom(s) 42, 111, 162, 186
customary law 68

dams 32, 103, 130
damages: climate, in DICE model 66; common law 56, 70
debt: ecological 130; traps in developing countries 50, 186
decentralization 56, 136, 229
decision-making 53, 85, 103, 109, 110, 145, 163, 184, 203, 229
decoupling ecological degradation from economic growth 24, 48, 49, 51, 177
décroissance see degrowth
deep ecology *see* ecology
deforestation 118, 121, 179, 215
degrowth: challenge 123, 230; contrasted with sustainable development 227–8; conferences 230, 231; core concepts of 97, 114, 227–31; definition of 6, 227–9; and democracy 229; and ecological economics 228; Group Assembly Process 231; and limits 228, 229; movement 6, 24, 161, 227–32; negative connotations of 228; origins of 228; policies 231; research related to 230, 231; as testing ground for ecological law 6, 217, 227–33
dematerialization 48, 49
democracy 111, 229, 230, 231
demography 36, 230
Department of Homeland Security 195
Depression *see* Great Depression
Descartes, R. 46
desertification 31
deterministic rules 109, 110
developed countries 120, 138, 176
developing countries 120, 138, 176, 186
development: agricultural 44; developmentalism 24, 45, 46, 47, 53; economic 187, 192; and expansion 3; human 29, 119, 120; of language 28, 30; policy 71, 175; pressures 65; resource 196, 201–2; societal 34; technological 15, 16, 28, 29, 30; *see also* Human Development Index; sustainable development
dialectics 102
DICE model 66
dilemma, human-Earth 1, 2, 11, 12, 14, 15, 16, 18, 22–7, 28, 29, 47–57, 65, 93, 127, 134, 138, 204
dioxins 118, 230
discounting 66, 145, 146
discourse 103, 144
disease 32, 35, 36, 183
distribution in ecological economics 22, 118, 119, 129, 135, 136, 137, 146, 230
diversity: bird 122; cultural 53, 130, 135, 228; and degrowth 231; genetic 156; knowledge 232; species 22, 31, 154; in systems 173; *see also* biodiversity
divine authority or supremacy 44, 45, 46, 101
Dominican Republic 139

doughnut economics *see* economics
Driver-Pressure-Impact-Response Framework 121
drivers, distribution of 116, 121, 122, 205
Dutch East India Company 49–50
dynamic systems 12, 25, 43

early modern era 17, 31, 32, 33, 35, 45, 46, 49, 50, 53, 185
Earth Charter 53, 95, 130, 152
Earth community *see* community
Earth Jurisprudence 97, 101–5
Earth Summit 152
Earth System Governance Project 141, 142
Earth systems: functioning of 2, 14, 101, 142; human relationship with 13, 28, 34, 37, 100, 156, 169, 229; life support capacity of 5, 57; science 28, 146
ecocentrism: human-inclusive 4, 6, 99, 104, 118, 128, 161, 172, 184, 227; in Oslo Manifesto 96; jurisprudence 99; and lens of ecological law 135; purist versions of 4, 98
eco-cultural restoration *see* restoration
ecological economics *see* economics
ecological footprint: based on consumption 120, 185; definition 131; and ecological overshoot 131; Global Footprint Network 124; and Human Development Index 120; and lock-in/lock-out assessments 6, 170; as metric of social metabolism 129, 131; and planetary boundaries 129; and metabolic rift 32, 185; research 122; of trade 70
ecology and ecological: capacity *see* capacity; debt 130; deep 52, 97; engineering 159; footprint *see* ecological footprint; functions 23, 67, 160; impacts *see* ecological impacts; indicators 95, 161; integrity *see* ecological integrity; justice 135, 172; limits *see* ecological limits; overshoot 119, 131; restoration *see* restoration
ecological impacts: of agriculture 31, 32; and the Anthropocene 12, 13, 15; control of 184, 205, 208, 210–11, 213–4; and economic growth insistence 132, 162; and efficient use of energy and matter 23, 24, 48; of fossil fuel use 34; and human population 35, 37; inadequate control of 54, 57, 178, 183, 188, 189, 204, 205, 209, 210–1; and individual rights and freedoms 173; and planetary boundaries 206; regional 115; and social metabolism 2, 130, 229; spatial and temporal reach of 2
ecological integrity: and attachment to place 152–63; contrasted with ecological health 154; as a core concept in ecological law 4, 95, 127, 138, 152, 177; definition of 99, 152–6, 163; and degrowth 228; discursive notion of 160, 161; and evolutionary trajectories 154–5, 163; flexibility of 155, 159; and historical human impacts 154–7, 163; in legal instruments 152, 163; and a mutually enhancing human-Earth relationship 97, 99, 105, 152; and planetary boundaries 157; practicable understanding of 156, 157, 158, 161; and precaution 112; and pristine nature 154; and restoration 5, 157–60, 175; socio-ecological integrity 99, 159, 180; and sustainability 94, 105; and trade 6, 188; in urban areas 155
ecological law: and attachment to place 152–64; and bee pollination 66; case studies 231–2; contrasted with contemporary environmental law 3, 65, 67; definition of 2, 93–6; and degrowth 6, 226–33; and ecological integrity 152–64; and environmental impact assessment 71; features of 3, 4, 127–47; and Indigenous legal traditions 161–2; and legal pluralism 95; lens of 231; and the Montreal Protocol 84; and a mutually enhancing human-Earth relationship 97–105; and naturehood 104; "ökologischen Recht" 93; and planetary boundaries 114–24; and private property rights 70; and sustainability 95; systems-based nature of 108–12; and trade 6, 69; as transformation of all areas of law 69, 232; transition to 5, 169–80
Ecological Law and Governance Association (ELGA) 96
ecological limits: and degrowth 228, 229, 230; and ecological law 2, 3, 67, 93, 95, 103, 127, 135, 136, 138–40, 177; and economic growth 49, 53, 69, 138; and environmental law 52, 53, 54, 86; and essential needs 134; and human flourishing 67; and limits-insistent narrative 11, 51; overshoot of 2, 3, 22, 24; and planetary boundaries 114, 115, 116, 118–22, 206; and precaution 142; precedence of technical and political

ecological limits *continued*
concerns 82; primacy of, in ecological law 4, 51–2, 129, 178; and resource productivity 133; and right to a healthy environment 104; and rights of nature 105; systems-based 129; and trade 187, 201, 204, 206, 208, 209, 210–16
eco-logics 162
economic growth: and capitalism 49–51; and climate modelling 66; correlation with ecological degradation 23, 24, 51, 98, 137; decoupling from ecological impacts 177, 183; and degrowth 228; and developing countries 138; and ecological law 134; and environmental backsliding 201; and environmental law 68; and European Union 136; infinite or perpetual 3, 4, 51, 54, 136, 162, 209; insistence on 11, 23, 24, 51, 98, 102, 108, 146; *Jobs and Economic Growth Act* (Canada) 194; and leverage points 172; and planetary boundaries 115; and sustainability 51, 94, 95; sustained 47–51, 57, 98; and trade 178, 185, 186, 189, 211, 212, 214; and United Nations policy 25, 48
economics: doughnut 119; ecological 129, 130, 228, 229; conventional 128, 130, 178, 214; heterodox 228; neo-classical 47, 66
economy: emerging 120; gift 31; globalized 32, 34, 37, 56, 69, 70, 118, 186, 201, 208; political 17; steady-state 97, 228
ecosphere 52, 67, 98, 129, 155, 212
ecosystem(s): alteration or disturbance of 12, 24, 29, 30, 31, 32, 33, 81, 82, 97, 98, 103, 112, 140, 153, 155, 156, 157, 163; -based criteria for environmental standards 73, 82; capacity of 31, 98, 130, 227; colonization of 131; complexity of 147; and complexity theory 109; contrasted with socially-constructed systems 110; and CPRs 160–1; diverse 4; downwind 83; ecosystem services *see* services; evolution and trajectory of 154, 159, 162, 175; functioning and dynamics of 51, 67, 98, 112, 141, 153, 155, 156; generally 2, 116; and HANPP 2017; health, in contrast to integrity 68, 154; human reliance on 30–3, 47, 77, 99, 112, 152; humans as part of 163, 177; hybrid 161; (integrated) global 22, 95, 99, 103, 117, 118, 230; interdependence with social systems 121, 173; management of 112; models 131; non-equilibrium nature of 4, 128; novel 159, 161, 163; pressures on 24, 28; pristine 153, 154, 155; protection 76, 78, 81, 83, 84, 152; receptor 80, 81, 82, 141; reference 159; remote, reliance on 32, 45, 185; restoration *see* restoration; and trade 183, 185, 211; understanding 127, 140; *see also* ecological integrity
eco-villages 229
Ecozoic era 99
Ecuador 103–4
Eden 98
education 36, 99, 111, 119, 230
efficiency: economic 118, 119, 129, 136; material and energy 23, 48, 79, 131–4, 178, 183
elders 162
electric utilities 82
elites and elite class 15, 17, 30, 44, 50
emergence in complex systems 12, 24, 108, 109, 110
employment 47, 146, 185, 231
endangerment finding 77, 78
energy: balances 118; and climate change 79, 178–9; conservation 100; decoupling 24; and degrowth 229, 231; demand reduction 51, 135, 137–8, 177; and doughnut economics 119; and ecological law 128, 129, 130–4; efficiency 178; extraction of 50; flow accounting 141; return on energy invested (EROEI) 34, 131, 147, 176, 179, 208, 215; fossil fuel(s) *see* fuel; and green building 70; -intensive luxuries 67; and IPAT 123; and law 111; and population 34, 35; in pre-industrial societies 31; nuclear 34, 79; solar 33; stocks and flows 5, 22, 28, 30, 46, 121, 123, 164, 170, 175, 178; and material throughput in the economy 2, 4, 6, 23, 24, 31–4, 47, 48, 49, 57, 229, 231; and trade 67, 118, 183, 189, 190, 204, 205, 206–8, 209, 211–15, 217; transition(s) 28–37; *see also* renewable energy
enforcement and enforceability: and climate change 179, 215; effective, definition of in NAAEC 198–9; environmental 53, 54, 189, 190, 196, 200–1; of law 42; and leverage points 211; of NAFTA and NAAEC environmental provisions 196; and

nested enterprises 161; regimes 54, 138; supranational 140, 187; weak 189, 190, 192, 196, 200; *see also* Commission for Environmental Cooperation, submissions on enforcement matters
Enlightenment 25, 46, 51, 57, 100
entitlement(s) 3, 42, 44, 55, 56, 70
entropic thrift 100, 210
environmental backsliding 191, 193–6, 211
environmental impact assessment (EIA) 68, 70–1, 127, 192, 194: strategic 71, 201; transboundary 190, 196–8
environmental law: anthropocentrism in 65–7, 128; as a boutique area of law 4, 53, 127, 130; Canadian 196; contrasted with ecological law 5, 65, 70, 127, 232; Ecuador 104; enforcement of *see* enforcement; European 140, 142; failures of 53, 65–71, 73–86, 183; generally 3, 4; global 68–9; holistic approaches in 52; and law as a complex adaptive systems 110, 177; and NAFTA and NAAEC 190, 191, 192, 193–5; piecemeal nature of 132; principles of 68; reductionism in 52, 67–71; rollbacks 193–5; stationarity in 174; US 52, 136, 139
Environmental Protection Agency (EPA), US 74, 75, 76–80
epistemology 109
epoch(s), geological 2, 11, 12–14, 163
erosion, soil 31
ethics: and the Anthropocene 15; and growth-insistence 23; and IPAT 123; and law 42, 43; as a normative regime 180; partnership 99
Ethiopia 160
Europe: common market 54, 139; European Community 54, 139; European Commission 143; European Court of Justice 54, 139, 143; European treaties *see* Treaty on European Union and Treaty on the Functioning of the European Union; European Union 54, 55, 73, 81–3, 135, 136, 139, 142–3, 172, 205; Europeans 16, 32, 45, 185
eutrophication 3, 33, 75, 81, 82
evidence: empirical 43, 103; legal 172, 202; scientific 23, 24, 140, 175
exceedance, of critical loads and levels 81–2
exchange, ecologically unequal 50
excludability 55
expansion and expansionism 3, 13, 24, 45, 46, 47, 49, 53
extinction, species 12, 13, 22, 29, 114, 117, 156, 206
extraction and extractivism 32, 34, 50, 131, 134–5, 178–9, 208, 214, 231

fairness: and growth insistence 23; and initial endowments 137; intergenerational, intragenerational and interspecies 4, 47, 103, 118–20, 122, 135, 136–8, 177, 230
famine 30, 35, 36
federalism 136
feedbacks, systems 1, 29, 35, 81, 140, 155, 169, 170, 173
fertility 36
fertilizers 32, 33, 50
finance and financialization 6, 15, 32, 49, 50, 57, 69, 111, 177, 178, 183, 184, 186, 188, 189, 204, 205, 209, 213–17
finiteness, of Earth 3, 5, 49, 51, 52, 119, 169
fire, human use of 14, 16, 28, 29–30
fisheries 34, 194, 200
Florida 160
flourishing 4, 67, 95, 98, 100, 122, 127, 137, 159, 229
food: crop production 207; demand 37, 208; and ecosystems 67; energy to produce 31, 35; as a human need 119; and metabolic rift 32; and population 97; producers and production 31, 36; sovereignty 230; storage 30; surplus 34, 44; systems 111, 142
footprint, ecological *see* ecological footprint
focal practice, ecological restoration and 159
foreign investment(s) 57, 70, 184, 185, 186, 210–11
forests and forestry 33, 34, 46, 50, 80, 93, 95, 98, 102, 115, 116, 118, 130, 140, 160, 162, 173
fossil fuels *see* fuel
freedom and freedoms: capabilities and 119; free movement of goods, services, capital and people 136; of information 172; in legal systems 1, 42, 44, 110; market 136, 139, 204; personal 46, 55–7, 67, 173, 204; as a principle of law 95; reproductive 36
French Revolution 43, 46, 56
freshwater 31, 33, 115, 116, 118, 140, 200

fuel(s): biofuels 36, 208; economy standards 77–8; fossil 2, 13, 14, 16, 17, 28, 29, 32, 33–5, 36, 121, 132, 177, 178, 185, 205
fundamental principles of law 95
future generations 4, 6, 23, 48, 52, 53, 54, 128, 130, 184, 191

G20, 48, 186
General Agreement on Tariffs and Trade (GATT) 187
General Agreement on Trade in Services (GATS) 187
gender 44, 119
geoengineering 16, 30, 180
ghost acres 31
gigantism 228
global ecological challenges 68, 190
global environmental law *see* environmental law
global heating 24, 66, 85
Global Reserve 139, 140, 216
globalization 13, 53, 68
gold 134–5
golden spike 13
goods: market 23, 31, 32, 47, 48, 66, 134, 136, 183–8, 190, 199, 204, 205, 206, 209, 211–2; non-market 57; public 55, 56; relational 231
governance: adaptive 146, 156, 161; analytical questions of 142; and climate change 176, 179, 205; of CPRs 157, 160–2; and degrowth 230; Earth system 141; and ecological law 97, 130; effective 131; environmental 68, 131; and IPAT 123; mechanisms 84, 129, 138; networks 1; as a normative discipline 15; novel and adaptive 16, 114; place-based 152; and planetary boundaries 120–3; and research and monitoring 139, 140–1; structure and infrastructure 3; and subsidiarity 135; sustainable 157; systems 53, 111, 157, 169; trade 205–6, 208, 215–16
government(s): and capitalism 49; centralized 56; and CPRs 176, 177; Ecuador 104; and EIA 71; and environmental violations 140; and global environmental law 68; and legal systems 111; and lock-in assessments 123; and markets 132; NAFTA 189, 190, 193–203; permits 74; and precaution 143; and private property 56, 70; and proportionality 136; regulation 176; and subsidiarity 55, 230; and support for limits-insistent narrative 24; US 52, 76; and taxation 54; and TMDLs 74, 75; and waiver of sovereign immunity 54, 139; world 229; and WTO 187
grain(s) 30, 36, 185
Great Acceleration 13, 130, 137
Great Depression 47
Great Jurisprudence 101, 102
Great Leap Forward 47
Green Belt Movement 160
green building 70, 133, 179
green economy 48, 187, 208
Green Economy Report, UNEP 208
Green Revolution 36
greenhouse gases: baseline 174; and Canada 194; and ecological footprint 131; and Ecuador 104; and growth 178, 213; and hydroelectric dams 38; intensity 133; locked-in 221; and ozone depletion 101; and planetary boundaries 121; and population growth 15; and socio-economic activity 130; and US 76–80, 86, 146
growth: economic *see* economic growth; -insistent narrative 4, 11, 23, 24, 48, 49, 51, 53, 54, 93, 103, 105, 114, 122, 171, 177, 178, 180, 183, 184, 209, 213, 227, 228; population *see* population
gross domestic product (GDP) 23, 34, 48, 49, 83, 120, 121, 122, 130, 131, 133, 134, 137, 231
gross national product (GNP) 48
guano 50

halons 84
Hanseatic League 185
happiness 23
Hardin, G. 55, 58
harm, ecological *see* ecological impacts
harmony with nature 152
heating, global *see* global heating
heavy metals 81
heterogeneity 35, 110, 116, 138, 162, 207, 212
history and historical: energy transitions 28–37; environmental 28, 29, 31, 119, 153, 156, 163; baselines in lock-in/lock-out assessments 174–5; fidelity 157; generally 3, 13, 17, 22, 24, 48, 49, 53, 102, 154, 171; trade 183–5
holism 96
Holocene 13, 14, 36, 155

home, single-family 134
homo sapiens 14, 35
honorable harvest 162
household 45, 134, 160
householding 100
human appropriation of net primary production (HANPP) *see* net primary production
Human Development Index (HDI) 120, 122, 124
human prospect 5, 12, 23, 97, 108, 157
human rights 46, 53, 96, 102, 103, 104, 130, 152
humanities 13, 15, 108
human services to ecosystems 100, 159
human-nature dualism 18, 153
hunter-gatherer societies 12, 29, 30
hunting 2, 3, 28, 29, 30, 35, 46
hydrochlorofluorocarbons (HCFCs) 85
hydrofluorocarbons (HFCs) 85
hydrological cycles 118, 206

Ice Age 36
Idle No More movement 232
impacts: anthropogenic 22, 25, 153, 154, 155; cumulative 52, 71; distribution of 118; ecological *see* ecological impacts; *see also* environmental impact assessment
impairment of water quality 75, 76
incentives and disincentives 121, 170, 171, 186, 195, 196, 199, 204
income(s): ceiling 230; countries 67, 204; individuals 119, 120, 229; inequality 173; tax 172; universal basic 231
India, Mughal 47
Indigenous peoples 32, 35
Indigenous worldviews 162
Indigenous legal traditions 152, 161–2
individualism 65
industrial agriculture *see* agriculture, industrial
Industrial Revolution generally 2, 31, 33, 35, 185; and the Anthropocene 12, 13, 14, 33, 155
industrialization 13, 28, 34, 132
industry 34, 83, 84
inequality 31, 43, 137, 173
innovation: generally 15, 36, 47, 146, 173, 186; in normative regimes 15, 180; policy 140, 201; technological 36, 48, 98, 102, 177, 183, 185, 229
institutions: environmental 65; European 139; financial 86; generally 42, 161, 173, 216, 229, 230; legal and political 57, 99, 110, 147; normative 19; social 99; tragic 203
instrumentalism 52, 129
integrated systems 141, 175
integrity: biological 153; biome 153; biosphere 153, 156, 207, 114, 116–17, 118; of the biotic community 153; of the commonwealth of life 153; ecological *see* ecological integrity; ecosystem *see* ecological integrity; in urban settings 155, 158
intelligence, artificial 30
intention 108, 109, 110, 157, 159, 163, 175
interdependence 95, 96, 100, 109, 110, 118, 121, 129, 176, 205, 209, 230
interdisciplinary approach 15, 217
intergenerational continuity 161
Intergovernmental Panel on Climate Change (IPCC) 78
international law(s): agreements 111, 152; generally 69, 111, 138, 232; and supranationality 178, 214; texts 98
interregional problem shifting 118
Inuit communities 118, 230
investment, foreign *see* foreign investment
investor dispute mechanisms 54, 57, 202, 203
IPAT 5, 7, 35, 123–4, 170
irrigation 31, 32, 45, 208

Jevons paradox 48, 131
Judeo-Christian worldview 56
judges 110, 111, 172
judicial authority and decisions 104, 110, 111, 139, 172
juridical persons 99, 111
jurisprudence: Earth *see* Earth jurisprudence
just compensation *see* compensation
justice: criminal 111; ecological 135, 172; economic 53, 130; generally 23, 46, 95, 119; intragenerational, intergenerational and interspecies 47, 96; social 51, 119

Khazzoom-Brookes Postulate 131
Kissimmee River 160
knowledge: diversity of 232; gaps in 140, 141, 146; human 109, 110, 142, 152; of place 162; scientific 74, 80; shared 143; systems 109, 135, 162; *see also* traditional ecological knowledge
Kyoto Protocol 176, 194

labor: cheap 50; conditions 50; and CPRs 160; and energy 31, 33, 35; factory 34; laws and standards 54, 93, 111, 232; and NAFTA 189; in poorer countries 37; productivity 23, 36; and property 56; and trade 216; and wealth 70
land: agricultural 32, 36, 118, 163; bioproductive 131; as commons 94; conquered 46; human domination of 24, 45; ice-free 98; irrigated 33; management 76; and property 56; public 158; and resource grabbing 32, 50; restoration of 158; -system change 115, 116; use 12, 31, 32, 37, 46, 93, 117, 118, 140, 141, 156, 159, 200, 206, 207, 208
landscapes 154, 155, 159, 160
language 3, 29–30, 65, 93, 228
law: administrative 93, 110, 111, 140; Anglo-American 70; codified or encoded 42, 44; contemporary 5, 100, 102, 104, 127, 162, 163; definition of 42–4; divine 101; ecological *see* ecological law; enforcement *see* enforcement; environmental *see* environmental law; great 101; natural 42–3, 101; positive 56, 70; sustainability 94, 95; *see also* Wild Law
lawyers 110
lead (Pb) 80, 81
legal codes 44
legal formalism 43
legal positivism 42, 43
legal realism 43
legal systems *see* systems, legal
legislators and legislatures 110, 111, 172
leverage points for systems change: generally 5, 164; in legal systems 169, 170–2; in lock-in/lock-out assessments 174–9, 184; and trade 209–17
life expectancy 34
lifestyles 30, 35, 94
limits: biophysical 94; on campaign expenditures 173; and degrowth 228–30, 231; ecological *see* ecological limits; energetic 32; to growth 114; on human activities 127; on human rights 103; -insistent narrative 23, 24, 93, 114–24, 129, 209, 227; living within 114; and mutually enhancing human-Earth relationship 97; regulatory 68, 78, 82, 83, 84, 121, 132, 136, 206; systems-based 73–4; on trade policies 187; *see also* planetary boundaries
linear thinking and linearity 25, 98, 108
load allocation in TMDLs 74
lock-in and lock-out 5, 6, 170, 171, 173, 184, 188, 203, 227
lock-in/lock-out assessment: applied to climate change 175–9; generally 5, 6, 164, 169, 231, 232; and planetary boundaries 122; and trade 209–17; use of 169, 174–5
Locke, J. 56, 204
Lone Pine Resources Inc. v. *Government of Canada* 201
low-carbon energy 70

macroeconomics 48, 51
Malthus, T. 36, 97
Mao 47
margin of safety 74, 116
markers, stratigraphic 12, 14, 16, 28
market: free 47; world 32, 50, 185; *see also* competition
material: ancient stores of 33; and energy flow accounting (MEFA) 129, 133, 141; and energy throughput *see* energy; extraction 4, 50, 128, 205; factors in the human-Earth dilemma 1, 2, 11, 14, 29; flows 5, 6, 22, 28, 46, 50, 67, 118, 123, 131, 164, 170, 175, 183, 189, 204, 205, 206–8, 209, 210–14, 217; and green building 133; growth 48; intensity 132; recycling and reuse of 70; use of 34, 37, 51, 100, 132, 134–5, 177; values 142
Matthai, W. 160
mechanization, in agriculture 24, 32
medicine 23, 34
medieval era 31, 49, 58
"merchants of doubt" 50
mercury 75, 76, 81, 84
Mesopotamia 44, 45
metabolic rift 31, 32, 185, 205, 208
metabolism: bioregional 31; industrial 69; social 2, 3, 22, 33, 35, 129, 131, 133, 134, 227; socio-economic 141
methyl bromide 1
methyl chloroform 1
Mexico: and DDT 189; and dioxins 118, 230; and trade 6, 54, 139, 189, 190, 192, 194, 195, 197
migration 36, 118
Migratory Birds Treaty Act 200
Mills, J.S. 97
Ming Dynasty 47
mining and minerals: early modern era 49, 35; and ecological law 134, 135, 232;

and fossil fuels 33, 34; metals 49; phosphate 33, 50; surface 195
models and modeling: economic 51; integrated 66, 208; environmental 80, 81, 131, 141
monarchs and monarchy 45, 46
money and monetization: abstractness of 66; and the Anthropocene 44; and consumption 67; and CPRs 160; and degrowth 231; and ecosystems services 56; and environmental law 65–7; and law 65, 102, 111; and mining 49; and the Neolithic Transition 31; and non-market goods and services 57; public 141
monitoring *see* research
monoculture 154
Montreal Protocol 54, 73, 84–6, 208
morality: and law 1, 42–3, 101; moral authority 45, 46; moral choice 14, 17; moral imperatives 102, 137; moral philosophy 46; more precepts 103; moral standing 100; moral truths 162; moral weight 23; right and wrong 42, 44
motor vehicles 76–8, 80
multilateral environmental agreements (MEAs) 69, 139, 191, 217
multinational corporations 50
mutually enhancing human-Earth relationship: and attachment to place 159, 161, 163; and degrowth 227, 228, 229; and ecological integrity 152, 154–7; and ecological law 4, 97–105, 114, 138; and law 11, 54, 86, 94, 108, 121; and planetary boundaries 123; as a societal goal 2, 4, 23, 108, 110, 119, 122; transition to 1, 5–6, 17, 123, 169, 170, 171, 172, 174–7, 180, 184, 187, 188, 189, 190, 201, 203, 204, 206, 208, 209, 210–17

nanotechnology 30, 34
narrative: growth-insistent *see* growth; limits-insistent *see* limits
nation states *see* States
National Ambient Air Quality Standards (NAAQSs) 79, 80, 81, 86
National Emission Ceilings Directive 88
National Environmental Policy Act (NEPA) 52, 70
National Highway Traffic Safety Administration (NHTSA) 77, 78
National Water Quality Inventory 75
nationalism 53
native species 161
natural areas 127
natural capital *see* capital
natural gas 34, 201
natural law *see* law
natural sciences *see* sciences
natural variability 153
nature: equilibrium view of 147; forces of 13; harmony with 52, 152; human 14, 15, 18; human domination and mastery of 98, 100, 163, 212; humans as part of 28, 163; as juridical person 111; laws of 94; market value of 135; and natural, meaning of 103; mutual flourishing in 159; nature-use 93; non-equilibrium view of 4, 147; non-human 25, 157, 162; protection of 93, 104; pristine 5, 99, 153, 155, 156, 157, 158, 163; respect for 53, 130; restoration of 157; rights of 101, 103–5; as separate from humans 18, 46, 52, 65, 96, 98, 100, 153; and society interactions 93, 131, 162; wild 98, 153, 157, 163
naturehood 104, 162
Nauru 50
needs, essential 134–5
Neolithic times 30, 156
Neolithic Transition *see* agriculture
nested enterprises 161
net primary production, human appropriation of (HANPP) 6, 37, 129, 141, 170, 206–8
nitrogen: fertilization 49; loading 117, 200; nitrates 50; nitrification 33; nitrogen dioxide 80; and planetary boundaries 75, 115, 116, 117; pollution 75, 81, 82, 83, 200
non-linearity 29, 43, 100, 109, 110, 115, 147, 175
nonpoint source water pollution 74, 75, 76
non-regression of environmental protection 191, 192, 193, 195–6
Nordhaus, W. 66, 71
norms and normativity: boundaries 14, 116, 229; choice 95; core of ecological law 2; desired 44; disciplines 15; foundation 98; frame 153; and governance 142; institutions 3; and law 65, 69, 102; regime, construct or system 4, 11, 14, 23, 47, 51, 57, 93, 99, 129, 180; political 42; response 16
North American Agreement on Environmental Cooperation (NAAEC) 54, 184, 189–90, 191–3, 195, 196–203; see also Commission for Environmental Cooperation

North American Free Trade Agreement (NAFTA): environmental provisions of 6, 184, 190–1, 192, 193–6, 200–3; generally 6, 69, 178, 184, 185, 188–90, 198, 200–4, 209, 212–4; Chapter 11, 54, 57, 139, 201–2, 214
Nova Scotia 201
novel ecosystems *see* ecosystems
novel entities 115
nuclear weapons 16
nutritional deficiency 22
nuisance 56, 70
Nunavut 230
nutrient: loading 24, 116, 118, 130; cycles 32, 140

Obama Administration 77, 78, 195
obligations 42, 44, 81, 82, 110, 179, 188, 189, 190, 191, 193, 194, 195, 196, 197, 215; see also responsibility
Occupy movement 232
ocean(s): acidification 115, 117, 119, 130; chemical composition of 12; dead zones 33; nutrient flows to 115; voyages 32, 185
Oceania 37
oil: exploitation of 104, 201; as a fossil fuel 34; palm 231; whale 33
ontology 109
open access resources 55, 164
Organization for Economic Cooperation and Development (OECD) 121
Oslo Manifesto for Ecological Law and Governance (2016) 65, 71, 95, 96
Ostrom, E. 160–1
overexploitation 55
owners and ownership: and ancient water law 45; and ecological law 232; and human domination of nature 25; of means of production 16, 17, 204; and property rights 56; remote 32, 50, 184, 211
ozone: stratospheric 51, 54, 114, 117, 118; -depleting substances 34, 73, 84–6, 208; ground-level 80, 81; hole 85; layer 3, 54, 86
Ozone Secretariat 85

Pacha Mama 103
paleoanthropocene 13
Paleolithic times 13, 17, 153, 156, 163
panarchy 1, 2, 5, 11, 24, 42, 43, 108, 109, 110, 111, 122, 155, 169, 174, 175, 184, 212
papal bulls 46
Paris Agreement 78, 79, 195
particulate matter 80
partnership ethics 100
path dependence 5, 109, 110, 155, 169, 171
PCBs 75
peace 46, 53, 70, 119, 130, 232
Pennsylvania 104, 105
permits 68, 74, 80, 83, 193
personal freedom *see* freedom
personhood, legal 104
Peru 50, 69, 139
pesticides 32, 208
phosphorus and phosphates: exhaustibility of 33, 34, 50; loading 33, 200; mining 50; and Nauru 50; and planetary boundaries 75, 115, 116; pollution 75
planetary boundaries: and *ACES* bill 79; and agriculture 33; and the Anthropocene 14, 17; and degrowth 230; and drivers of environmental change 52; and ecological law 3, 4, 10, 95, 102, 112, 115–18, 127, 128, 129, 131, 137, 138, 141–2, 144, 145, 146; and ecological integrity 153, 156–7; and environmental law 73, 75, 78; and fairness 118–20; and fossil fuels 35; and governance 16, 114, 120–3, 138, 139, 140, 141–2, 144, 145, 146; and IPAT 123–4; and limits-insistent narrative 24, 97, 114–24; and lock-in/lock-out assessment 5, 170, 175–6, 177, 178–9; non-negotiability of 115; researchers 85, 114; and trade 183, 184, 187, 189, 206, 207, 213–17
pluralism, legal 42, 95, 110, 111, 136, 231
politics and political power 9, 50, 176
pollination 66, 67
polluter pays principle 68, 127
pollution, control of 23, 73, 132
population, human: centers of 185; growth of 2, 13, 24, 44, 52, 97, 131, 132, 134, 208; as material factor in human-Earth dilemma 29, 35–7; overpopulation 36; trends 205; *see also* IPAT
positive law *see* law
poverty 22, 23, 45, 137, 160
power: distribution 15; dynamics 47; fossil fuel 17, 33–4, 185; and governance 142; human 28, 31; imbalance 17, 46, 47, 55, 203; of monarchies 46; networks and structures 1, 45, 46, 111, 138, 227;

plants 51, 76, 78, 83, 84, 195; political 50, 103, 108, 195; relationships 17, 31; solar 2; sustained 17; water 34
precaution and precautionary principle: and catastrophic harm 144, 146; and the Earth Charter 152; and ecological law 4, 112, 128, 142–6, 163, 174, 177; and economic growth 132; and environmental law 68, 127; and incommensurability 146; and irreversible loss 145–6; and leverage points 210; and lock-in/lock-out assessment 215; and ozone depletion 85; and planetary boundaries 115, 116, 140, 156, 206; and rights of nature 104
preferences, human 24, 65, 98, 102, 115, 136
primacy: ecological 135; of ecological integrity 155; of ecological limits 4, 52, 93, 95, 105, 114, 127, 129, 177, 179, 209, 210, 211, 215, 216; of economic growth 47, 54, 98, 102, 178, 213
principle of sustainability 94, 95
private property *see* property
privateers 46
privatization 136, 160
productivity: generally 75, 231; labor 23, 36; resource 24, 48, 132, 133, 178, 213; *see also* net primary production
profit(s): and capitalism 47–50; and growth-insistence 23, 57, 103; and mining 134; and NAFTA Chapter 11, 201, 202; and research 141; and trade and investment generally 37, 178, 185, 186, 189, 204
prohibitions 1, 42, 44, 110
property: common 111, 179, 210, 215; private 49, 56, 111, 185, 209, 230; public 111, 179, 210, 215
property rights: and contemporary law 47, 55–7, 67, 70, 99, 108, 136, 173, 184; and ecological law 70, 96; and the Enlightenment 46; and freedom 204; and growth-insistence 93, 98; and interconnectedness of life 43; and trade 178, 187, 204, 212, 214, 215
proportionality 4, 55, 128, 135–6, 139, 143
protectionism 69, 139, 186, 188
public opinion 50
public participation 70, 191, 194, 197, 202, 203
pulp and paper mills 200

Qing Dynasty 47

Quebec 122, 201–2

race 44
"race to the bottom" in trade 189, 193
radiative forcing 78, 114
radionuclides 13, 16
reason and rationality 43, 51, 65, 66
rebound effect 48, 131, 132, 133, 134
reciprocity 162, 197
recycling 70, 133
reductionism 25, 43, 51–3, 65, 67–71, 104, 108, 183
regeneration 4, 103, 104, 112, 128, 152
region(s): regional challenges 200; regional context 158; regional economies 124, 186; regional impacts 37, 115, 118, 122, 129, 205; regional level or scale 16, 55, 114, 116, 121, 135, 140, 161, 172, 178, 185, 209, 212, 214; regional limits 230; regional trade 185, 186, 188, 190, 203, 204
regulation(s): climate 76–9, 116; and CPRs 160; and ecological law 4, 128, 232; environmental 56, 65, 69, 70, 73, 74, 84, 85, 139, 189, 190, 191–2, 195, 200, 201; and Indigenous legal traditions 162; and legal systems 111; and leverage points 172; limited 184, 188, 212; local 58; resistance to 57; over- 188; trade 69
relationship, right 23, 97, 119, 122, 153; *see also* mutually enhancing human-Earth relationship
religion 42, 44, 48, 163
remote owners and investors 32, 183–4, 211
renewable energy 31, 33, 70, 79, 133, 178, 179, 208
reproductive freedom 36
"republic of science" paradigm 143
research: Anthropocene 15, 17; and degrowth 6, 227; and ecological law 4, 6, 128, 140–2, 146; and EIA 71; governance-related 139, 212; planetary boundaries 34, 78, 114–18, 122, 129, 130–1, 137, 156, 157
Research & Degrowth 142
resilience: and adaptiveness, compared 173; definition of 171–2; ecological 171; engineering 171; and planetary boundaries 119; and restoration 158, 159; as systems feature 5, 110, 169, 170, 171; and Transition Towns 232; value-neutral nature of 169

responsibilities 45, 70, 96, 111, 122, 205
restoration: definition of 157; ecological 112, 152, 157–60, 161, 163, 175; and ecological law 4, 128; eco-cultural 5, 152, 157; and human choice and intention 157–8; reciprocal 158
right relationship *see* relationship
right(s): constitutional 70, 104, 105, 215; economic 105; of nature *see* nature; social 105; to a healthy environment 104; to develop 120
Rio Declaration on Environment and Development (1992) 48, 142, 152, 196; *see also* Earth Summit
Rio Principles 25, 152
Rio+20 Summit 48, 152, 187
risk: assessment 143; contrasted with uncertainty 143; management 143
rivalness 55
rivers 44, 75
Rockström, J. 114
royalties 54
rulemaking ruts 203
Russia, Czarist 47
Rwanda 160

salinization, soil 31, 45
sanctions, graduated 160
scale: local to global 3, 5, 16, 102, 111, 114, 121, 153, 155, 163, 184, 209, 233; multi-scale 100, 161; spatial 1, 2, 6, 11, 35, 52, 103, 108, 131, 155, 157, 161, 174, 206, 210, 212, 217; temporal 1, 2, 6, 11, 35, 52, 103, 108, 155, 157, 161, 174, 206, 210, 217
scarcity 133
science: natural 14, 16, 29, 102; physical 108; scientific reasoning 46; scientific consensus 50, 85; social 13, 15, 16, 17, 29, 108; sustainability 131, 140
sea level rise 116, 205
Security and Prosperity Partnership 197
self-organization 110, 154
Sen, A. 119
services: ecosystem 56, 66, 100, 146, 158, 159, 160; free movement of 136; goods and 23, 32, 47, 48, 57, 66, 136, 183, 184, 185, 186, 187, 190, 199, 204, 205, 206, 209, 211, 212; public 231
shareholders 47, 134, 204
shared resources 45
Shrimp-Turtle case 188
slavery and slave trade 32, 33, 43, 45
silver 49
Small Is Beautiful 228
smallpox 32, 35
social constructs 53, 55, 57, 93, 109, 112, 119
social hierarchy 43
social metabolism *see* metabolism
social preferences 65
social sciences *see* science(s)
social status 30
society and societies: agricultural 2, 12, 17, 29, 33; capitalist 16; civil 111, 213; collapse 45; complex 45; global 53, 66, 130; goals 4, 98, 110, 118, 119, 120, 180; hunter-gatherer 12, 29, 30, 31; industrial 17, 32, 94, 132, 185; modern 44, 46; and nature 93, 131; professional 24; settled 44; stratification of 17, 44; Western 94
Society of Ecological Restoration 158
source-receptor relationship 76
sovereign immunity 54, 139, 201
sovereignty: and ecological law 96; and European exploration 46; as factor in human dilemma 53–4, 57; food 230; and growth-insistence 93, 98; and supranational authority 138–9; and system lock-in 173; and trade 184, 185, 211
Soviet Union and Soviet Bloc 47, 93
spaceship Earth 52
speculation, financial 49, 134
St. Lawrence River 201
Stockholm Resilience Centre 114
storage, food, grain or commodities 28, 30
stratification, social 30, 31, 44, 55
stratigraphy 11, 12, 14
stratospheric ozone *see* ozone
strong sustainability *see* sustainability
subsidiarity 4, 55, 128, 135–6, 139, 177, 230
sulfur dioxide 73, 80, 81, 82–4
supranational authority 4, 46, 53–5, 128, 138–40, 177, 178, 187, 188, 211, 214, 216
surplus production 44
sustainability: science *see* science; strong 97, 119; weak 50, 94
sustainable development: Sustainable Development Goals (SDGs)
sufficiency 23, 137, 211
Sweden 120, 138, 206
symbiosis 157
synthetic chemicals 34
syphilis 32

systems: adaptiveness *see* adaptiveness; aquatic 34; -based approach 1, 2, 16, 17, 51, 52, 112, 163, 209, 210; change 144; complex adaptive *see* complex adaptive systems; cultural 36, 44; dynamic 12, 25, 43; economic 56, 70, 136, 142, 173, 179, 183; ecosystems *see* ecosystem; feudal 46, 49; industrial 45; initial conditions 97, 109, 169; institutional 11, 171, 174; multi-scalar 161; normative 14, 47, 51; political 3, 45, 48; socio-ecological 98, 109, 122, 132, 141, 173, 174, 175, 206, 208; thresholds 24, 78, 86, 95, 112, 115, 116, 118, 140; *see also* emergence; path dependence

taxation and taxes 54, 111, 133, 172, 177, 231
technological optimism *see* technology
technology: and the Anthropocene 12, 14, 15, 16; breakthroughs 115; carbon-neutral 132; energy-saving 133; and energy transitions 28, 29–30, 35, 36; innovation 48, 98, 102, 183, 185, 229; and IPAT 23, 35, 123–4; low-impact 230; optimism 24, 49, 98, 177; zero-emissions 132
TEEB Report 66
thermodynamics 51, 94, 95, 183
Thomas of Aquinas 101
tipping points 35
Tobacco Strategy 50
Total Maximum Daily Load (TMDL) 73–6, 80, 81, 86
tourism 130
toxic chemicals and pollutants 34, 84, 95
trade: agreements 57, 69, 70, 194, 185, 186, 189, 201, 202, 203, 204, 205; dispute settlement 54, 57, 187, 203; and ecological law 183–217; flows 67, 187; global system of 16, 22, 69, 178, 183, 185, 188, 190, 203, 204, 209, 214; history of 45, 184–5; liberalization 69, 185; lock-in assessment of 183–217; networks 45, 185; *see also* North American Free Trade Agreement, World Trade Organization
tradeable pollution rights 84
trade-offs 51, 145, 206
Trade-Related Aspects of Intellectual Property Rights (TRIPS) 187
traditional ecological knowledge 163
Tragedy of the Commons 6, 55, 58
tragic institutions 6, 184, 203, 212
transboundary impacts 196–8
transdisciplinary approach 1, 12, 15, 16, 17, 232
Transition Towns 6, 232
transportation 34, 79, 111, 133, 186
travel 45, 79, 118
Treaty of Rome 54, 139
Treaty on European Union (TEU) 55, 135
Treaty on the Functioning of the European Union (TFEU) 135, 142–3
trespass 56, 70
tropical forests *see* forests
Trump Administration 78, 196
Tuna-Dolphin case 188

uncertainty: and ecological law 4, 128; irreducible 12, 98, 101, 103, 108, 112, 155, 175; and precaution 142–5; zones of 114–15, 116, 175
unintended consequences 98, 109, 112
United Nations: United Nations Framework Convention on Climate Change (UNFCCC) 78; United Nations Economic Commission for Europe 80; United Nations Environment Programme (UNEP) 48, 208
United States: border with Mexico 189, 195; Constitution *see* United States Constitution; environmental backsliding in 195–6; Supreme Court 56, 70, 71, 76, 77, 173
United States Constitution 43, 70, 139, 172, 173, 214
United States—Mexico—Canada Agreement (USMCA) 6, 184, 185, 188, 189, 201, 203, 204, 209
universe 101, 102
urbanization 7, 28, 34, 52, 98, 111, 162, 207
utility and utilitarianism 4, 51, 65, 108, 128, 134, 136, 137, 179

value(s) and valuation: and democracy 229; ecological 146; environmental 56, 188; hierarchy 47, 69; monetary 65, 66, 163; and planetary boundaries 24, 115; shift 134
Vilcabamba River 104
violators of law 140
voters 110

waste: assimilation capacity 133; dump 95; and ecological footprint 131; flows 6; loads 133; processing 132, 133; production of 131, 134, 135, 137; reduction 208

wasteload allocation 74
water: availability 118; blue 115; demand 70; fossil 34; generally 2, 5, 24, 31, 34, 77, 93, 94, 133; as a human need 119; irrigation 45; law 45; management 32, 160; pollution 31, 74, 75, 195; quality 73–5, 80, 93; supplies 74
weak sustainability *see* sustainability
wealth: accumulation 50, 228; distribution 36, 137; inequality 3, 15, 17, 173; maximization 23, 65, 67, 204; restraint on 135; sufficient 134, 137; virtual 49
wedge analysis 133, 147
well-being: community 152; ecological 129, 231; generally 6, 22, 23, 65, 99, 136, 146, 187, 191, 232; human 116, 117, 119, 137, 228; social 188
Western knowledge systems 162
Western law 162
Western thought and worldviews 46, 52, 98, 163
Westphalia, Peace of 46, 54
wetlands 31, 75, 84, 194
Wild Law 97, 101–3, 172
wilderness 32, 153
wildlife 29, 74, 77, 93
wind 2
wood 33, 34, 36
worldview: anthropocentric 104; Cartesian 65; ecocentric 104, 128; and ecological law 96; generally 43, 108, 111, 170, 172; Indigenous 162; mechanistic 25; Western 98

Yasuni National Park 104
Yellowstone National Park 160

For Product Safety Concerns and Information please contact our EU representative GPSR@taylorandfrancis.com
Taylor & Francis Verlag GmbH, Kaufingerstraße 24, 80331 München, Germany

www.ingramcontent.com/pod-product-compliance
Ingram Content Group UK Ltd.
Pitfield, Milton Keynes, MK11 3LW, UK
UKHW020011200926
475702UK00014B/1828

* 9 7 8 0 3 6 7 6 1 2 9 5 5 *